GEOMETRY BY CONSTRUCTION

GEOMETRY BY CONSTRUCTION

Object Creation and Problem-solving in Euclidean and Non-Euclidean Geometries

MICHAEL MCDANIEL

Universal-Publishers
Boca Raton

Geometry by Construction:
Object Creation and Problem-solving in Euclidean and Non-Euclidean Geometries

Universal-Publishers
Boca Raton, Florida
USA • 2015

ISBN-10: 1-62734-028-9
ISBN-13: 978-1-62734-028-1

www.universal-publishers.com

Cover image: Kamira/Bigstock.com

Publisher's Cataloging-in-Publication Data

McDaniel, Michael.
Geometry by construction : object creation and problem-solving in euclidean and non-euclidean geometries / Michael McDaniel.
pages cm
Includes bibliographical references and index.
ISBN: 978-1-62734-028-1 (pbk.)
1. Euclid's Elements. 2. Geometry, Non-Euclidean. 3. Geometry, Modern. 4. Geometry—Foundations. 5. Geometry—Problems, Exercises, etc. I. Title.
QA445 .M34 2015
516—dc23
                                        2015930041

Contents

Michael McDaniel
Aquinas College

To Dennison and Marguerite Mohler and to Ted Thompson: your vision has inspired new and interesting geometry. Thank you for the opportunities!

Preface

In 1919, a solar eclipse allowed British researchers to photograph stars whose light had to pass the sun before reaching Earth. Twelve years earlier, Einstein had predicted the sun's mass would bend the fabric of space, making such a star appear out of place in the sky. He had even provided the formulas to determine how much the star's position would be misplaced. When the photographic plates were developed and measurements taken, the star matched his prediction and the universe became more clearly understood. The planets are not billiard balls, spinning and rotating around each other in a box, their movements governed by their gravitational attraction; the box is more like a trampoline and the stars and planets bend the trampoline.

As a consequence, we know the universe does not fit Euclidean geometry. The geometry which has served us so well on Earth all these years is not exactly correct at the celestial level. Although we cannot hitch rides with angels and perceive our universe from some point outside it, we can play with models of bent geometries, which is exactly what we will do in this book. In fact, we will look down on the bent spaces and see them in their entireties. We will be outside the bent spaces, working in a Euclidean space which contains these bent spaces.

Using only the most ancient of drawing tools, the compass and straightedge, we will make the two-dimensional objects which might illustrate the essential structure of the universe. For those of us with access to a dynamic geometry program, like *Geometer's SketchPad*, we can bring the ancient tools into our century of animations, precise measurements and color printing.

The only way to fully appreciate the geometrical worlds ahead is with a notebook, compass and straightedge handy. Prepare to keep your hands busy.

On the philosophy of the book

We focus on constructions for some good reasons. We get to make and see what we are studying. We also experience geometry as a living discipline because this book contains some new results from undergraduate research and a few new ideas from the author which tie the content together. For example, this book contains the constructions for hyperbolic lines and elliptic lines which are a single reflection away from being the same construction. Some texts have results from living mathematicians; here we have new, verified results written when the authors were students.

In this way, the text has been written as a door to a hallway to another, more important door: here is an introduction to actual, publishable undergraduate research. Down the hall of undergraduate research, we find the satisfactions and challenges of full-time mathematical work. It is the author's greatest hope that this book helps a few more students through those doors.

Throughout the development of the theorems and constructions in this book, the student contributors and the author noticed that every time proof was sought for a solid conjecture, there was always enough material to demonstrate the truth. We waded into the unknown, heading toward the desired result and every time, what we needed existed. The search put professor and students on the same footing: none of us knew how to find the answer! We developed confidence and comfort in the generosity of mathematics. Beyond the richness of the subject, we often felt as some philosophers have written, the Truth was there, waiting to be discovered. More than waiting – an active give and take, sparring, wrestling took place as we sought the support we needed for the theorems.

Compare this experience with the mundane, disappointingly common pastime of playing Solitaire. How many times do we go through the deck, never finding one of the cards we need to continue the game? The majority of times, we lose. But in math, when we seek the truth, we almost always find what we need to prove that truth. In fact, when a problem eludes proof for a long time or ends up as one of the unprovable ideas, that problem gets famous!

Whether the reader believes this or not, our advice might be worth taking. When the Truth is talking, listen and take advantage. Trust that you are strong enough for the intellectual battles and that the math itself wants to help! The successful reader will find the unwanted terms cancel out, or the angles we need congruent lie in similar triangles, or whatever it takes to force the true conclusion; the Truth is in reach.

Geometry by Construction

Chapter 1

Euclidean geometry rules and constructions

Since our two-dimensional versions of bent space require Euclidean geometry, we will start with that geometry. Here's a brief review of the geometry of Pythagoras, Newton and Kant: Euclidean geometry. When the time comes, we will see that the bent geometries have the same logical standing as Euclidean. In fact, we will study the bent spaces in terms of Euclidean, sort of like astronauts who work in orbit: some Earthly rules apply while others do not apply.

These pages of rules and constructions are absolutely essential knowledge for this book. We will explore many of those rules. We will also refer to rules as if they are familiar, even if they have not turned up in any previous problems.

1.1   Euclidean Geometry   Vocabulary and Definitions

Learn these quickly and never get them mixed up.

Point: That which has no size; it has zero dimensions. A point is pure location.

Line: The set of points which lie between two distinct points and the points included when the segment is extended. When the segment is extended in only one direction, this makes a ray. Lines, segments and rays are one dimensional. Points on a single line are called collinear.

Plane: The set of all points on two intersecting lines, along with the points between any two points, one on each line. A plane is two dimensional. Points in the same plane are called coplanar.

Between: Point B lies between points A and C if the distance from A to B plus the distance from B to C equals the distance from A to C.

Isometries

An isometry is a transformation which preserves distances and angle measures.

Reflection: Reflect across a line. The line acts as a mirror. The reflection sends each point to the point on the line perpendicular to the line of reflection, at the same distance from the line of reflection as the starting point. Usage: *Ref l* (object) where l is the line of reflection and object is the object being reflected. This is the only basic isometry which reverses orientation.

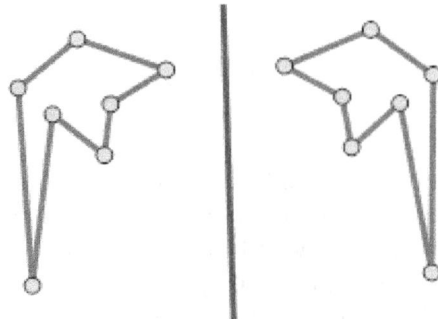

Translation: Move the object in the plane. All points have their x and y coordinates changed by the same two amounts. Usage: $f(x, y) = (a + x, b + y)$.

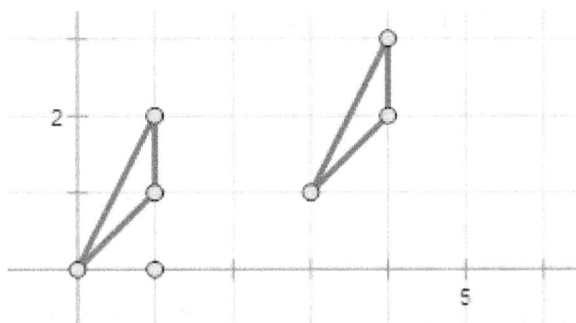

Rotation: Rotate an object around a point by an angle of a certain size. Usage: $\text{Rot}(P, \theta,\text{object})$ where P is the point around which the rotation occurs and θ is the angle of rotation.

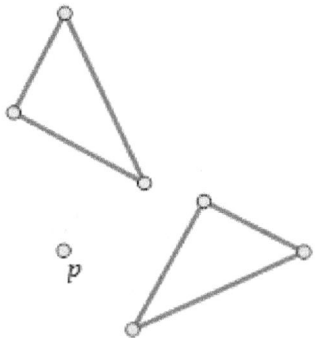

Special Lines

Notation: We will use the nouns segment, ray and line before two letters to indicate which object we mean. The length of the segment, AB, has no modifier and this indicates a number.

Parallel: Coplanar lines which have no intersection are parallel.

Perpendicular: Lines which meet at right angles are perpendicular.

Segment Bisector: A line containing the midpoint of a segment.

Angle Bisector: A line through the vertex of an angle which divides the angle into two congruent angles.

Angle: Two rays, a ray and a segment or two segments joined at an endpoint (called the vertex) form a set of points called an angle. We define 360 degrees as the rotation of one ray around its end point exactly once; so one degree is 1/360 of a complete rotation.

Special Angles

These angles get their names from their positions.

Adjacent: Two angles with the same vertex and one shared side, but with no interior points in

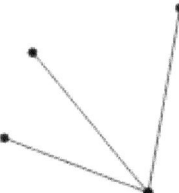

common.

Vertical: Two intersecting lines form two pairs of equal angles. Each pair is vertical. They form

a letter X-shape.

The following special angle positions occur when two lines are crossed by a

transversal.

Corresponding: This pair are in the same relative position, as in both being upper right. A pair of these form an F shape. The two lines cut by the transversal are

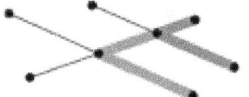

parallel if and only if corresponding angles are congruent.

Alternate Interior: Their sides form a Z-shape. The two lines cut by the transversal are parallel if and only if the alternate interior angles are equal.

Alternate Exterior: These angles are vertical with a pair of alternate interior angles.

The two lines cut by the transversal are parallel if and only if the alternate

exterior angles are equal.

Same-side Interior: These angles form a block C-shape.

The two lines cut by the transversal are parallel if and only if the same-side

interior angles are supplementary.

Same-side Exterior: These angles are adjacent to a pair of same-side interior angles.

The two lines cut by the transversal are parallel if and only if the same-side

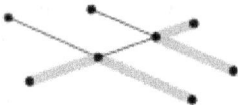

exterior angles are supplementary.

Supplementary: Two angles are supplementary when their angle measures add up to 180 degrees.

A special case of supplementary angles: Adjacent angles with collinear exterior sides are supplementary. These are sometimes called a linear pair.

Complementary: Two angles are complementary when their angle measures add up to 90 degrees.

A special case of complementary angles: Adjacent angles with perpendicular exterior sides are complementary.

External Angle: When one side of a polygon is extended, the angle formed by the extended side and next side of the polygon is an external angle. In a triangle, an external angle is equal to the sum of the remote interior angles (the two interior

angles which are not adjacent to the exterior angle.)

Straight Angle: An angle whose measure is 180 degrees is a straight angle. The exterior sides form a line.

Obtuse Angle: An angle whose measure is between 90 and 180 degrees.

Right angle: An angle whose measure is 90 degrees is a right angle.

Acute Angle: An angle whose measure is less than 90 degrees is an acute angle.

Bisectors

Perpendicular Bisector of a segment: A line, ray, or segment which is perpendicular to a segment at that segment's midpoint is a perpendicular bisector of a segment.

Angle Bisector: A line which cuts an angle into two equal angles is an angle bisector. Any point on an angle bisector is equidistant from the sides of the angle.

Circles

A circle is the set of all points equidistant from a center point.

Radius: A segment from the center to any point on the circle. All radii are congruent.

Chord: A segment whose endpoints are on a circle.

Diameter: The largest chord in a circle. A chord which contains the center.

Arc: The part of a circle between two points. We name arcs greater than or equal to 180 degrees with three letters.

Inscribed angle: An angle with its vertex on the circle and whose two sides are chords or a chord and a tangent. An inscribed angle equals half the measure of its

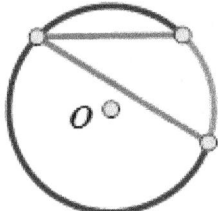

arc.

Tangent: A line which intersects a circle in exactly one point. The radius is perpendicular to the line at the point of tangency.

Secant: A line which intersects a circle in two points.

Central angle: An angle whose vertex is the center of the circle. The central angle

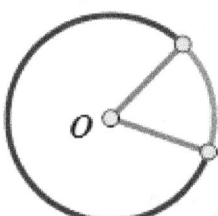

has a measure equal to its arc.

Sword Theorem: A diameter which is perpendicular to a chord bisects the chord and its two arcs. Alternatively, if a chord is the perpendicular bisector of another chord, then the bisector is a diameter. Also, if a diameter bisects a chord, then it

is perpendicular to the chord.

Concentric circles: Concentric circles have the same center and different radii lengths.

Triangles

A triangle consists of three non-collinear points (vertices) and the segments joining them.

Degenerate: If the three vertices are collinear, then the triangle is actually just a segment. Or all 3 points are the same point. Each case is referred to as a degenerate triangle.

Isosceles: An isosceles triangle has two sides congruent. The third side is called the base and the vertex opposite the base is The Vertex.

Acute: All angles in an acute triangle are less than 90 degrees.

Right: A right triangle has an angle of 90 degrees.

Obtuse: An obtuse triangle has an angle of more than 90 degrees.

Equilateral or **Equiangular:** A regular triangle: all sides and all angles are congruent.

Median: The segment from a vertex of a triangle to the midpoint of the opposite side is a median.

Altitude: The segment from a vertex perpendicular to the opposite side is an altitude.

Similar: If two pairs of angles of two triangles are congruent then the triangles are similar. Similar triangles have corresponding sides in proportion. So if triangle ABC is similar to triangle DEF, then $\dfrac{AB}{BC} = \dfrac{DE}{EF}$. An alternative method for showing two triangles are similar is by having a proportion of two pairs of sides and having the included angles congruent.

Congruent: A pair of congruent triangles have the same size and the same shape. Five ways to prove two triangles are congruent have acronyms based on the corresponding parts which match up.

   SSS: If two triangles have three pairs of congruent sides, then the triangles must be congruent.

   SAS: If two triangles have two pairs of congruent sides and the included angles are also congruent, the triangles are congruent.

   ASA: If two triangles have two pairs of congruent angles and the included sides are congruent, the triangles are congruent.

   AAS: If two triangles have two pairs of congruent angles and a corresponding pair of non-included sides are congruent, then the triangles are congruent.

   HL: If two right triangles have a pair of sides (legs) congruent and their hypotenuses are congruent, the triangles are congruent.

These five ways to prove triangles congruent are theorems, not axioms. The proofs rely on isometries.

The Sine Law: For any triangle ABC, the following proportions hold.
$$\frac{\sin A}{a} = \frac{\sin B}{b} = \frac{\sin C}{c}.$$

The Law of Cosines: For any triangle ABC, the following equation holds.
$$a^2 = b^2 + c^2 - 2bc \cos A.$$

Quadrilaterals

Convex: If a band is stretched around the convex quadrilateral, the band sticks to the sides with no gaps. Precisely, the segment between any two points on or in the

quadrilateral is in the quadrilateral.

Concave: If a band is stretched around the concave quadrilateral, there is a place where the band does not touch the quadrilateral. Precisely, there exist two points of the quadrilateral such that the segment between them contains points outside

the quadrilateral.

Cyclic: If a circle can contain all four vertices, the quadrilateral is cyclic.

Trapezoid: A quadrilateral with only one pair of opposite sides parallel is a trapezoid.

Square: A regular quadrilateral: all angles and sides are congruent.

Rhombus: An equilateral quadrilateral.

Rectangle: An equiangular quadrilateral.

Parallelogram: Opposite sides parallel.

Kite: A quadrilateral with two pairs of consecutive, congruent sides is a kite.

1.2   Constructions to know

The constructions which follow are for compass and straightedge. (*SketchPad* does these with a few clicks.) Some colored pens or pencils are useful for complicated drawings. Always place your drawing page on a stack of paper: this gives the point of your compass a place to dig in and stabilize. Also, choose a compass without a rivet holding the pivot together. A nut and bolt or screw can be tightened. (As you use your compass, the arms will get a little loose.) Being able to tighten your compass keeps your drawings accurate. Some compasses end with a narrow cylinder, able to hold only a standard pencil. By choosing a compass with an adjustable barrel, you can switch pens and pencils in order to keep your drawing organized by color.

When constructing, we almost never get to pick points out of thin air. One instance where we do get to start from nothing is beginning the construction. Even when we are given a length or angle to use, we will almost always have to copy these to our own work space. The initial placement of the first segment requires a bit of forethought: leave space for your work to happen. Do not begin near an edge of the paper and then build so that your work quickly lies outside the margin. Assume that your construction will take up a lot of space, so give yourself room and make your starting objects large enough to fit all the stuff you plan to do.

Copying Lengths and Angles

To copy a length, lay down a segment which is longer than the one you intend to copy. (The given segment is unlabeled.) Then place the point of your compass at one end of the given segment and adjust the radius so that your pencil tip meets the other endpoint. Make a little arc. Now move your compass to the workspace and place the compass point at one endpoint of your long segment (point A.) Mark an arc with your pencil on the segment. The arc intersects the segment at a point. Segment AC is congruent to given the segment.

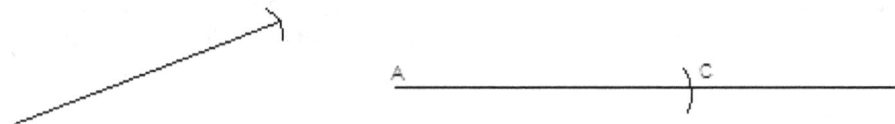

To copy an angle, again draw a segment DE in your workspace. Mark an arc across the angle you wish to copy and then put the point of your compass at endpoint D to mark an arc with the same radius, across the segment in your workspace. Go back to the first drawing and place the point of the compass at a point where the arc crosses a side of the angle. Adjust the radius so that the pencil lands on the other intersection of the arc with a side of the angle. You have just captured the width of a chord PL, which will help build a congruent angle. Mark a little arc in order to show what you have done.

Place your compass point at the intersection of the arc and segment in your workplace M. Draw an arc QR which intersects your first arc. Construct a segment through their intersection and the endpoint of the segment which you have used before, DK. The first segment and this segment form an angle congruent to the given angle.

It is usually the case that, to prove a construction works, we will have to draw segments which are otherwise invisible in our construction. It will also be good to remember that copied radii are congruent, even when the entire circles are not drawn.

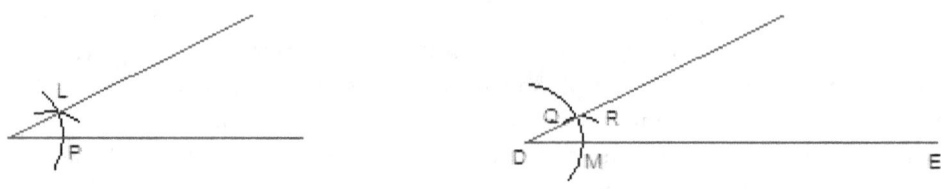

On paper, you must also leave behind marks which indicate how a construction was performed. Keep the extras to a minimum: do not draw arcs much longer then they need to be. Place construction marks well away from where the action is whenever possible, leaving room for future steps. A good construction avoids needless complexities like extra points of intersection.

Bisectors

We have two kinds of bisectors: angle bisectors and segment bisectors. In both cases a line does the actual bisecting: the bisector is named after what gets bisected. It might be said that the midpoint of a segment bisects the segment; but we will not say that here because the midpoint is a point on the segment; a bisector goes beyond the object being cut in half.

The angle bisector construction starts similar to copying an angle. This time, we draw arcs from both points where the first arc intersects the sides. The point N where the arcs intersect combines with the vertex of the given angle to define the angle bisector.

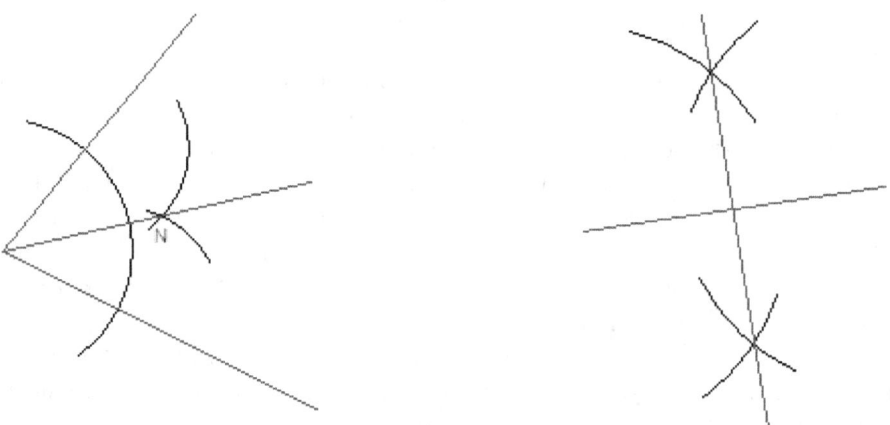

The perpendicular bisector of a segment is one of the easiest constructions we have. Simply choose the radius of your compass greater than half the length of the given segment. Place the point of the compass at each endpoint and draw the arcs long enough to intersect above and below the segment. The perpendicular bisector passes through the two points where the arcs intersect.

Perpendicular Segments

There are two more situations where we will construct specific perpendiculars: through a point on a line and through a point not on a line. Their constructions are virtually the same, so we will show the construction for a point not on a line. We are given the point C and the segment AB. Since the point C was so close to one endpoint, we extend the segment (the dashed part) in order to get some working room. The reader who has been doing the constructions all along can probably see that we made an arc which cut the extended segment twice. Then we used those two points of intersection as centers for two new arcs which intersected on the other side of the segment. The undrawn segment CJ is perpendicular to segment AB and it contains C, as required.

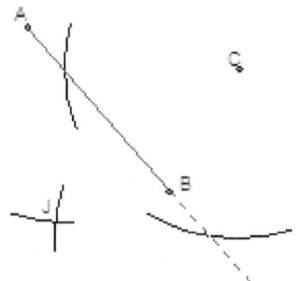

All the geometry we will do in this section is called Euclidean geometry because of its connections with the Greek mathematician Euclid. His Elements were volumes of geometrical knowledge which set the example for mathematical exploration: the author gives the assumptions and definitions and builds from there using logic. Many books exist which deduce geometry in exactly this way: they follow the precise reasoning which handles every detail of the development of Euclidean geometry. This work establishes geometry in ways which few subjects outside mathematics can match – assumptions made clear, all results are known to be true, no editorial agenda, no quest for big bucks.

We, however, will follow an easier path than a purely formal development. We will indeed follow the rules of logic and prove many of our theorems, but some results we will simply take as true and move on. We will rely on more than axioms, definitions and previous conclusions. We will construct the objects of our study using compass and straightedge so that we can see what we are talking about.

Everything we make in this course is made of points. A point has no size; it is pure location. The first Euclidean axiom assumes that two distinct points determine exactly one line. This is our first construction.

Write down two dots, the points A and B. Now take your ruler and draw the unique line segment through those two dots. If you write an arrow on the ends of the segment, you have constructed a line. We won't be drawing those little arrows at the ends very often. But we will have occasions where the accuracy of the notation allows us to explain our work with great precision. This notation is not meant to drive everybody crazy with pickiness. But there are times when we need to refer to a segment, a ray or a line so that every reader knows what we mean. So here are examples. We have pictured segment $\overline{AB}$, ray $\overrightarrow{CD}$ and line $\overleftrightarrow{EF}$. It would be incorrect to call the ray $\overrightarrow{DC}$ because, for the ray, we say the endpoint letter gets written first and endpoint letter gets the endpoint symbol.

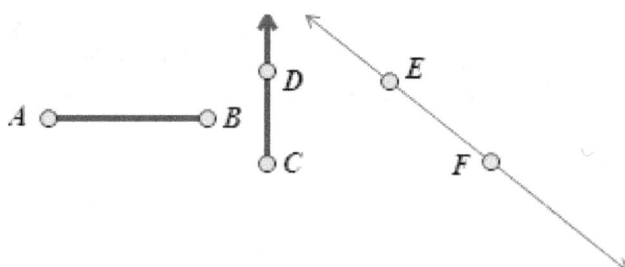

This is the first of many notations to keep clear. The letters AB alone stand for the length, which is a number. The segment $\overline{AB}$ is a set of points, not a number. Such attention to detail, however, could get tedious in a rapid conversation. On those occasions, we will allow AB to stand for whatever it needs to stand for and, if the writing establishes context, that meaning should be clear. It is hoped this double standard resembles real life: everyday-talk is quick and easy, written work requires attention to detail.

The two arrows on the ends of a line stand for another Euclidean axiom – lines can be extended indefinitely. This is a useful assumption when two lines are constructed which do not meet on the paper; yet, theoretically, we have reason to believe they do in fact meet eventually. Then we don't have to tape another piece of paper to our first one in order to imagine their intersection.

The figures in this book were made in *SketchPad*. The lines there do not have the arrows drawn in; but the arrows do appear when lines are printed as part of a drawing. Constructing the line through two points is a crucial skill. One way is to click on the two points and go to the Construct menu. As long as only those two points are selected, the choices construct segment, construct line and construct ray will be enabled. The order in which the points were selected determines the way the ray points, just the same as the way we write them down. When two lines are constructed so that they intersect, select the two lines and construct intersection. *SketchPad* then extends the drawing so that the point of intersection exists in the drawing. The reader might have to scroll to see it; but the point of intersection is there.

The first axiom matches the way we construct: we very rarely get to draw a line just anywhere. Usually, we have two significant points present in the drawing; these determine a line. Then we place a pencil-tip at one point, hold the ruler next to the pencil and along the second point and we draw the desired line. As easy as it sounds, some care must be taken to get the line through the two points, not just near them.

Two points also determine a circle when one point is the center and the other is a point on the circle. (The existence of circles is the next Euclidean axiom.) In *SketchPad*, select the points in the order we want and the choice Construct circle by center and point is enabled under the Construct menu. Drawing a circle on paper is what a compass is built to do. First, get a few sheets of paper under your construction so your compass point has a foothold. Place the point at the center and put a little pressure on it as you draw the circle. Draw the circle once, nice and neat. Drawing it twice only makes the circle thicker and if the compass wiggles a little bit, the correct circle is lost in a whirl of errant arcs.

1.3   Tangent construction

Suppose we wanted to construct the light rays going from the sun to the Earth. Compared to the sun, the earth is a speck. So we'll treat the Earth as point A and the sun as circle O. We are going to need tangents to circle O through point A because the rays going from the sun to Earth are the only rays we want. Any other light rays cannot or will not hit point A. We are not allowed to simply lay down the straightedge so that we draw a line through A that appears tangent to circle O. We must find a point of tangency and lay down the edge along A and the point of tangency: two points determine a line. Precision will get us safely through the bent spaces and that precision starts now. Stated as a geometry question, we have the following given.

Problem. Given a circle with center O and a point A outside the circle, construct the two lines through A tangent to the circle.

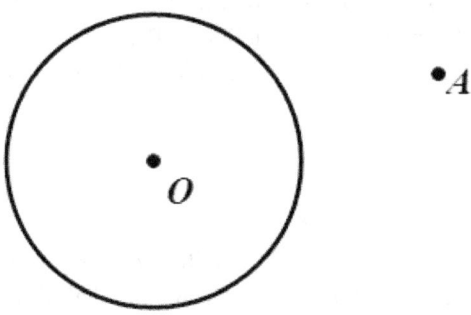

We only have two specific points. We could make a line, ray or segment through them. We could use one as a center and the other as a point on a circle in order to draw another circle. In other words, we do not have a huge list of options. Yet, choosing wisely requires strategy, not guessing. A useful strategy is to know each word in the problem. What do we know about lines tangent to circles? The radius drawn to the point of tangency forms a right angle with the tangent line.

Here is where a thorough knowledge of the basics gives us a huge boost. An angle with its vertex on the circle, one of whose sides is a radius and the other is a tangent line is an inscribed angle. An inscribed angle equals half its arc. The inscribed angle we want is a 90 degree angle, which means it can be inscribed in a semicircle. The reader is advised to stop here and think. How can we use the above knowledge to perform this construction?

Proof: Construct $\overline{OA}$. Then construct its midpoint M using the perpendicular bisector (compass and straightedge) or by selecting the segment in *SketchPad*. Then construct the circle with center M and radius $\overline{MA}$. This new circle intersects the given circle in two points, D and E. These are exactly the two points we need. Here's why. The undrawn angle ADO is inscribed in a semi-circle. Therefore it must be 90 degrees. One side of the angle is the radius $\overline{OD}$. The other side is the line $\overleftrightarrow{AD}$. Since $\overleftrightarrow{AD}$ is perpendicular to the radius at the point of intersection, the point of intersection must be the point of tangency. So $\overleftrightarrow{AD}$ is tangent to the given circle. The same argument holds for the line $\overleftrightarrow{AE}$.   ■

The proof relied on a property of inscribed angles: an inscribed angle has measure equal to half its arc. Reviewing the Euclidean Rules is worth doing every once in a while. The reader should keep all the basic ideas fresh so we can recall topics when we need them.

1.4   Cyclic quadrilaterals

Since we just used inscribed angles equal to half their arcs, let's use this fact to explore a family of special quadrilaterals. Cyclic quadrilaterals have their four vertices on a circle. We will now prove the theorem that a quadrilateral is cyclic if and only if its opposite angles are supplementary.

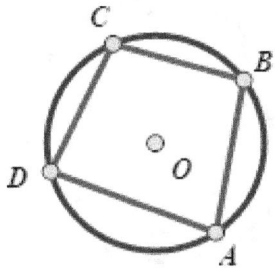

Cyclic quadrilateral
ABCD.

An "if and only if " compound statement needs two proofs. We take one of its component statements as given and prove the other statement follows. Then we switch their places and prove what is often called, "the other direction." So, for starters, we will work with the quadrilateral *ABCD* being given as cyclic. So the figure above fits the situation.

The alert math student has probably seen some useful structure in the figure: all the angles are inscribed angles. We just got done considering inscribed angles in detail. Using a recent idea in the current problem happens a lot in math texts.

A quadrilateral *ABCD* is cyclic if and only if its opposite angles are supplementary.

Proof: The key idea is to notice that the arcs cut by angles *B* and *D* form the entire circle. So the sum of these two angles is half the sum of degrees of a circle. Half of 360° is 180°, so we know that angles *B* and *D* must be supplementary. Angles *A* and *C* get the same consideration. This finishes the first proof.

The second proof says that if the opposite angles of a quadrilateral are supplementary, then the quadrilateral is cyclic. This time we have to pretend that we don't know the four vertices lie on a circle. So the first figure is inappropriate for the second proof. Our second figure will use such a drawing and we will use a proof by contradiction. We will assume that the opposite angles of quadrilateral *EFGH* are supplementary but the four points do not lie on a circle. Now, any three non-collinear points lie on a circle; so we have a circle through points *G*, *H*, *E* but not through *F*. We have placed *F* outside this circle when it could have been inside. That version of the proof is similar to the one we are about to do.

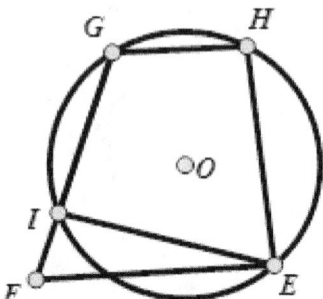

The point I occurs where $\overline{GF}$ intersects the circle. From the first proof, we know that angles H and GIE are supplementary. Since angle F is also supplementary to angle H, angles F and GIE are equal! This forces angle IEF to be zero, which forces F and I to be the same point, contradicting our assumption that F lies outside the circle. The contradiction shows that our assumption is false and that proves the other direction. ■

1.5 Similar triangles

If we draw the diagonals of a cyclic quadrilateral and hide the quadrilateral, we can find a pair of similar triangles on a circle as pictured.

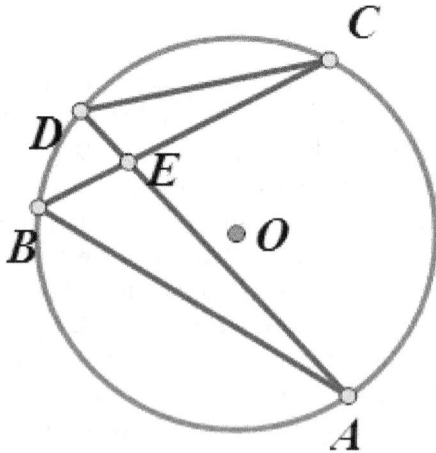

Similar triangles.

The most common way to prove two triangles are similar is to show that they have two pairs of angles equal. This is easily accomplished in the figure above since inscribed angles A and C cut the same arc BD and inscribed angles B and D cut the same arc AC. So triangles ABE and CDE are the same shape but different sizes. Please note that the order of the letters used in naming the triangles illustrates their corresponding parts. This is more than a mathematical nicety; math people expect this sort of precise notation.

Similar triangles, along with congruent triangles, form the most useful building blocks for planar geometry proofs. From the earliest grade school figures to complicated drawings beyond college, triangles are always around. Let's take a look at an important theorem whose proof requires the use of similar triangles.

The theorem of Menelaus is as old as it sounds. He worked in Alexandria around A. D. 100 when the library there housed a legendary compendium of human intellectual achievement. Although we will not consider this theorem later, it is worth noting that this also holds in spherical geometry.

1.6   The theorem of Menelaus

Given three points D,E and F, one on each (possibly extended) side of triangle ABC; the three points D, E, F are collinear if and only if the product of their ratios

$$\frac{AF}{FB} \frac{BD}{DC} \frac{CE}{EA} = 1.$$

Furthermore, if an orientation is applied to the triangle ABC, an odd number of the segments involved in the calculation will be measured against the given orientation.

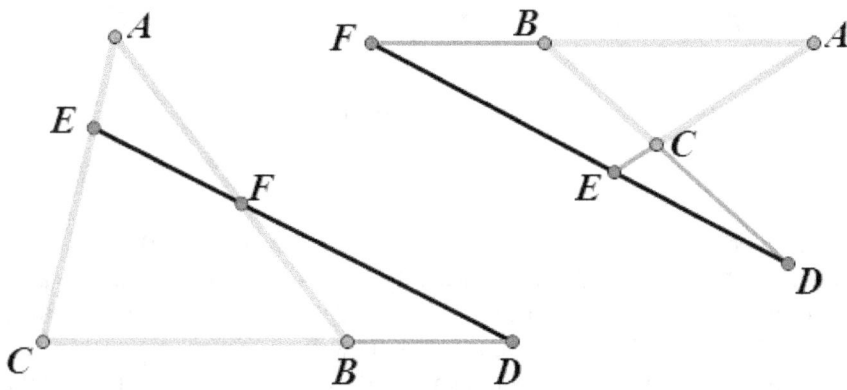

Two formations of Menelaus for $\triangle ABC$.

The theorem is an " if and only if " which means we have two proofs to do. Its proofs illustrate the use of ingenuity – knowing similar triangles are involved does not lead to a quick explanation. In fact, none of the triangles in the figure illustrating the theorem have to be similar. Before we start the proof, let's get to know the theorem. Use the example above and notice how the lengths may change but the product of fractions does not.

If the points D, E, and F are collinear (in the first drawing above), then the segment lengths have interdependency – we can't just pick numbers for all of them. Let's assign lengths $6 = AF$, $9 = FB$, $x = BD$, $12 = BC$, $8 = CE$ and $y = AE$. Substituting into Menelaus's theorem gives us $\dfrac{16x}{3xy + 36y} = 1$. If we solve this for y, we see that $y = \dfrac{16x}{3x + 36}$. So when $x = 1$, $y = \dfrac{16}{39}$.

Now imagine the point F is like a pivot, so that the point D can get further and further away from B. That would be like taking the limit as x approaches infinity. Then the length y has a limit of $\dfrac{16}{3}$. Although this is a sort of calculus idea, the geometry supports the concept: $\overline{AE}$ has a size limitation because so many other segments have fixed size. Only EF, FD and BD are changing in our teeter-totter vision.

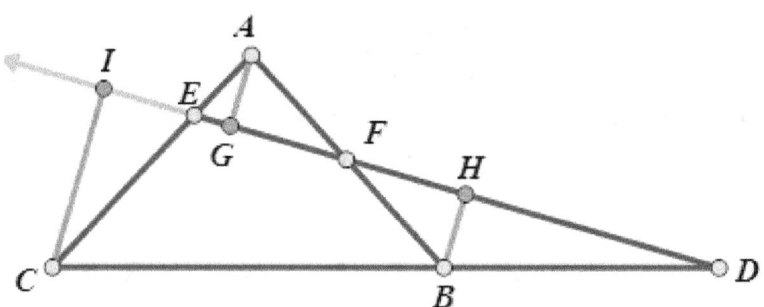

For the proof of the theorem.

Proof.   The theorem has an " if and only if. " Let's start with the case which fits the drawing: D, E, F are given collinear. We will keep our goal, the equation using the product of three ratios, in mind as we build this proof. The conclusion we seek often guides our initial choices, in math as well as life.

If the points D, E, and F are collinear, we have to prove that the product of ratios $\dfrac{AF}{FB}\dfrac{BD}{DC}\dfrac{CE}{EA} = 1$. A close look at the above figure shows three new segments which were not in the original figure. These new segments each start at a vertex of the original triangle and are perpendicular to the line through D, E, F. Such perpendiculars are always possible because Menelaus's theorem never uses an entire side of the original triangle as one of its segments. (It is possible that the line through D, E, F is already perpendicular to one of the sides of the triangle. This causes no major difficulty, so we will visit that version in a homework problem.)

These three new segments are the sort of contrivance which frustrates some students. They wonder who thinks of these things and the obvious answer is, Menelaus. They wonder how students are supposed to think of these things by themselves and that is precisely the soul of this book. We gain expertise from experience. If our cleverness kicks in once in a while, bonus.

We have to look at what we want: a product of ratios. Then we think of sources of ratios: only similar triangles comes to mind. Using perpendiculars creates angles which are equal, which eases our search for useful pairs of triangles.

Note the structure of the choices. We don't just drop perpendiculars at random. In this case, we have used each vertex of the original triangle once and each perpendicular was sent to the same line. When we name the similar triangles, we could use three pairs of triangles, where each triangle uses one of the new segments. If we write our ratios in a clever way, perhaps these new segment lengths will cancel, leaving us the lengths we care about. This is in fact what happens.

We can prove our first pair of similar triangles by noting $\angle CIE = \angle AGE = 90°$ and that angles IEC and AEG are vertical angles. Those give $\triangle AEG \sim \triangle CEI$, which implies $\dfrac{AE}{AG} = \dfrac{CE}{CI}$. Using the same strategy at each new segment, we get $\triangle AFG \sim \triangle BFH$, which implies $\dfrac{AG}{AF} = \dfrac{BH}{BF}$ and, $\triangle DIC \sim \triangle DHB$, which implies $\dfrac{CD}{IC} = \dfrac{BD}{HB}$.

Consider the three equations we just wrote with a fraction equal to a fraction. Multiplying the left-hand sides together and the right hand sides together maintains the equality. We also have some lengths in common (AG, CI, BH) and these lengths are in position to cancel! A little algebra leads to the desired equation. Our first half of the proof is done already. This illustrates a theme throughout the book: the problem contains enough information. The fractions we needed were available, with some extra stuff. When we went after exactly what we needed, the extra stuff canceled out!

The second half of the proof starts with the product of ratios true and we have to show this forces D, E, and F to be collinear. We are going to assume that the three points are not collinear. Since two points determine a line, we will use the line through E and D and label its intersection with side $\overline{AB}$ as G. The next figure illustrates our situation.

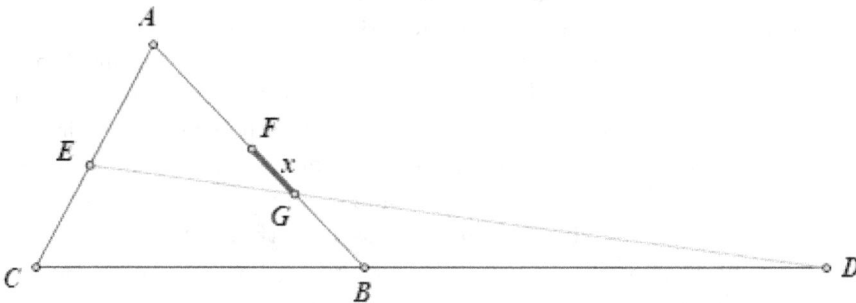

We have marked the distance $FG = x$. We need to take a moment to discuss a subtlety of this theorem. The theorem is often stated with the product of ratios set equal to -1 because, if orientation is considered, an odd number of the lengths will go against the orientation of the $\triangle ABC$, making the product negative. In this proof, we are working to establish the lengths only: we will not be concerned with minus signs. Each segment length is non-negative, as well as the unknown length x. Also, we have assumed F is between G and A, when it could just have easily landed between G and B. This second possibility would not change the structure of our proof. So we will continue using this figure to stand for both cases.

Remember, we are given $\dfrac{AF}{FB}\dfrac{BD}{DC}\dfrac{CE}{EA} = 1$. The line through D, G, and E satisfies the first part of the proof, which gives us $\dfrac{AG}{GB}\dfrac{BD}{DC}\dfrac{CE}{EA} = 1$. A little algebra leads to the equation

$$(AF)\,(GB) = 2x + x(GB + AF) + (AF)(GB).$$

This forces x to be zero. (A negative x is impossible because x is a length.) If $x = 0$, the points F and G are the same, which contradicts their being different. So our assumption leads to a contradiction, as required.   ∎

1.7   sAs for similar triangles

Two triangles can be proved similar if we can get two pairs of angles equal. (Since the sum of the angles of any triangle is 180 degrees, the third angles must be equal as well, giving us two triangles of the same shape.) We will now establish a second method of showing triangles are similar.

If $\triangle ABC$ and $\triangle DEF$ have angle $\angle A = \angle D$ and the proportion $\dfrac{AB}{AC} = \dfrac{DE}{DF}$, then the triangles are similar.

Proof: We will suppose the triangles are not similar. This means they are not of the same shape. So we can move the smaller triangle inside the larger one by isometries, lining up the corresponding sides as pictured. The two sides $\overline{EF}$ and $\overline{BC}$ are not parallel, since the triangles are not similar. So we have drawn the ray $\overrightarrow{EF}$ intersecting the ray $\overrightarrow{BC}$ at point G. The attentive reader knows what will happen next. (Given two such triangles, we could construct such an example ourselves by copying lengths.)

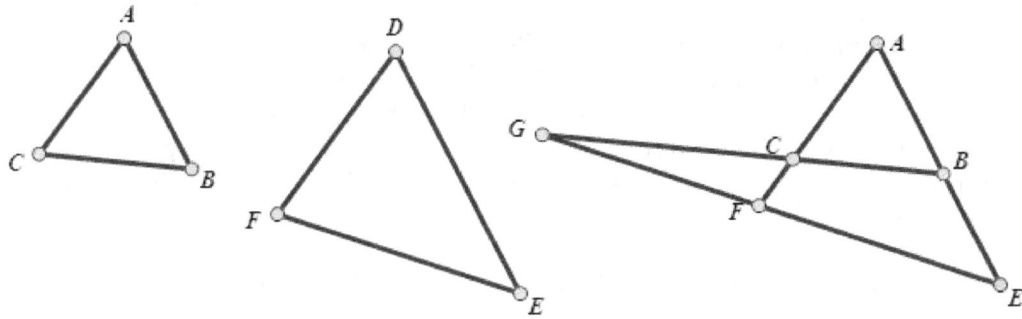

That last figure has the Menelaus theorem's structure. Therefore, we have the product of ratios $\dfrac{BE}{EA}\dfrac{AF}{FC}\dfrac{CG}{GB} = 1$.

We can rewrite $CG = CB + BG$ in that equation, then split up across the numerator to get the equation

$$\frac{BE}{EA}\frac{AF}{FC}\frac{CB}{GB} + \frac{BE}{EA}\frac{AF}{FC}\frac{GB}{GB} = 1$$

With a little algebra, we can show that our given proportion $\dfrac{AB}{AC} = \dfrac{DE}{DF}$ is logically equivalent to $\dfrac{BE}{EA}\dfrac{AF}{FC} = 1$. This gives us the second term of the equation equal to 1, which forces the product $\dfrac{BE}{EA}\dfrac{AF}{FC}\dfrac{CB}{GB} = 0$. That is our desired contradiction. So we have the triangles similar and the corresponding sides $\overline{EF}$ and $\overline{BC}$ must be parallel.   ∎

It is worth mentioning that the relative positions of these sides and angles is the SAS formation, the same as the way we prove two triangles are congruent. In summary, we have two ways to get triangles similar now: AA and sAs. This funny-looking sAs stands for the version where we use parts of sides in proportion, like $\dfrac{BE}{EA}\dfrac{AF}{FC} = 1$, as used in the proof.

We have five ways to prove triangles are congruent: SSS, SAS, ASA, AAS and HL. Since AAA would only guarantee similar triangles, there is really only one more combination of the letters S and A missing from the list: SSA. Of course, HL is a version of SSA. But it is worth noting that SSA is not a method of proving triangles congruent. In fact, it is possible to choose lengths and angles in the SSA configuration such that no triangle is possible, one triangle is possible and two triangles are possible. It is also a relief that, because of the vulgarity associated with a shift of these letters, SSA is not one of the five ways to prove triangles congruent.

1.8   Ceva's Theorem

Ceva's theorem is often found next to Menelaus's theorem in geometry textbooks with good reason: the product of ratios looks the same. The pattern is simple: start at a vertex of the triangle and choose a direction. Follow that direction and write down vertex, point; point, vertex; vertex, point; point, vertex; vertex, point; point, vertex. Those are the twelve letters of both theorems. Ceva's theorem even sounds like Menelaus's.

Given $\triangle ABC$ with points D, E, F one on each side, the Cevians $\overline{AD}$, $\overline{BE}$ and $\overline{CF}$ are concurrent if and only if $\dfrac{AF}{FB}\dfrac{BD}{DC}\dfrac{CE}{EA} = 1$.

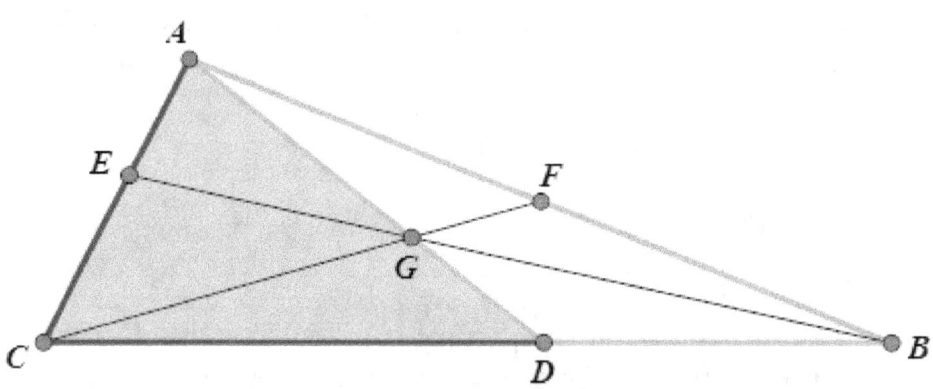

Ceva's theorem with Cevians.

   A segment from a vertex to another side of the triangle is called a Cevian. The altitudes, medians and angle bisectors of a triangle are famous Cevians. Concurrent segments intersect together at one point.

   Proving Ceva's theorem is not as daunting as it looks. The figure above has been provided a shading scheme which will help with the proof. The reader should take a moment to think: what would be a good strategy?

Perhaps the way we tackled Menelaus's theorem would work again. That theorem also had an " if and only if. " We started with the collinear points given and we were able to find some useful ratios through similar triangles. The other direction was a proof by contradiction which had the support of the collinear points leading to the equation. Let's begin in the same way: we work with the Cevians concurrent and we will try to prove the equation is true. That means we can work with the drawing above. Now, where can we get some useful ratios?

While similar triangles might work, there is a very fast proof using two applications of Menelaus's theorem. Each Cevian can be considered as the line cutting three sides of a triangle in Menelaus's theorem. That means we have three products of ratios available! Keep the goal in mind: the product of ratios we seek does not fit any of these three options because the product we seek does not use any parts of any Cevians. How do we decide what to use?

One option is to write down all three possibilities and look for connections. What if we could find products with some factors in common? Maybe we could multiply them together and some factors would cancel, leaving just what we need. That is, in fact, exactly the right idea.

Proof: We will apply Menelaus's theorem to $\triangle ABD$ and $\triangle ADC$. We get
$\dfrac{AF}{FB}\dfrac{BC}{CD}\dfrac{DG}{GA} = 1$ and $\dfrac{AG}{GD}\dfrac{DB}{BC}\dfrac{CE}{EA} = 1.$
Multiplying left-hand sides and simplifying gives the desired product of ratios. The right-hand sides multiply to 1. That finishes the first part of the proof.

The second part is very similar to the second part of the Menelaus proof. The reader should try working through this half of the proof using the following outline. We now have the product of ratios given and we have to prove the Cevians are concurrent. By supposing they are not concurrent, we can draw in a Cevian which does share the same point of intersection as the other two. Its endpoint will have to be different from the point on that side used in the given information. This gives us an application of the first part of Ceva's theorem and another product of ratios. The same trick we used in the Menelaus proof works again: call the difference between these two points on the same side x and rewrite the equations using this substitution. Simplification will force $x = 0$.   ■

There is a version of Ceva's theorem which uses angles inside the triangle instead of segment lengths, sometimes called Trig Ceva. Its statement and proof is in the homework as well as other assertions we have made.

1.9   Incircles, excircles and circumcircles

Some important constructions use the famous Cevians: angle bisectors, medians and altitudes. Every triangle has an incircle, the circle inside the triangle, tangent to each side. The center of the incircle is called the incenter and it is the intersection of the angle bisectors. If the interior and exterior angles of the triangle are bisected, their intersections form three excenters, the centers of the three excircles. An excircle is tangent to each side of the triangle, but it lies on the outside of the triangle.

Constructing the incircle and the excircles is easy in *SketchPad* and difficult by hand. This is because a little error in one angle bisector can put the incenter way out of position. In *SketchPad*, we click on the vertices of angle in the order we would write the angle's name, then use the angle bisector command in the construct menu. By hand, we have to construct two angle bisectors as carefully as we can. In either case, finding the incenter or excenter does not end the work. We will have to construct a perpendicular from the center to a side in order to determine the radius.

Here triangle ABC has two excircles and an incircle.

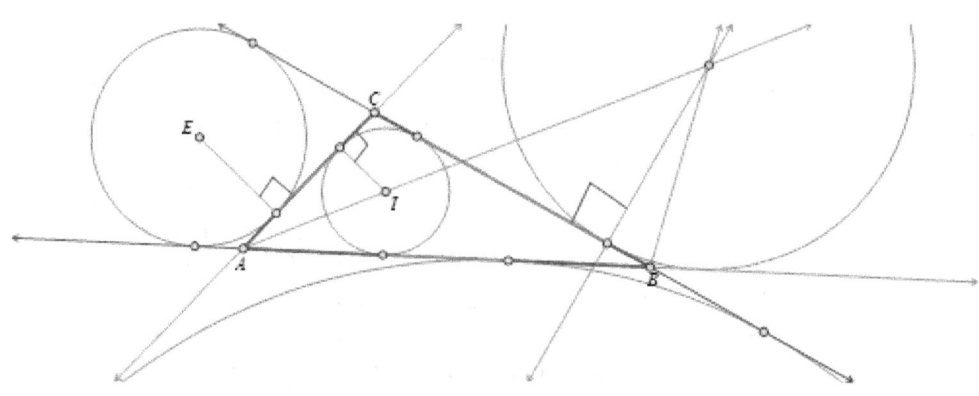

If we construct the circle with center on $\overline{AC}$ using the points of tangency from the excircle and incircle as endpoints of a diameter, we get circle O which turns out to be an interesting example after we study inversion across a circle.

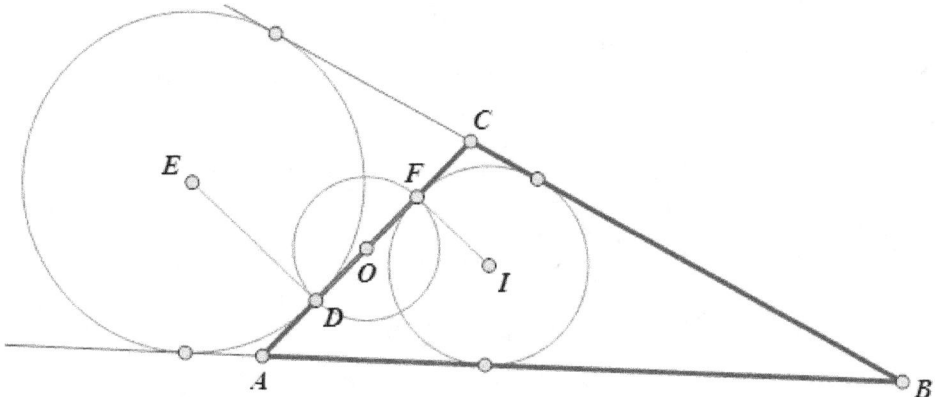

Incircle, excircle and circle O.

Every triangle has a centroid, its balancing point. Finding the coordinates of the centroid of a triangular region is a calculus problem. We can easily construct the centroid of any triangle by constructing two medians. To construct a median, find the midpoint of a side using the perpendicular bisector construction. Then just draw the segment from the opposite vertex to the midpoint. The triangle in the figure below has its centroid, G, constructed. We use two perpendicular bisectors of sides to find the circumcenter, O, the center of the circle through the three vertices. This is illustrated in the triangle in the figure below. Note that perpendicular bisectors do not contain Cevians because they do not necessarily pass through a vertex. The intersection of two altitudes is the orthocenter, H. All three altitudes are concurrent at H. The letter names G, H and O are pretty much standardized in geometrical literature.

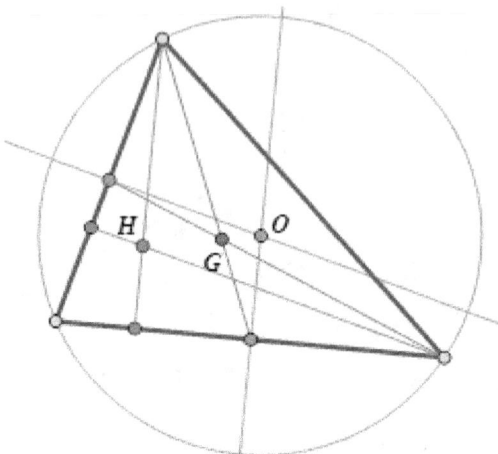

Centers H, G and O.

1.10   The 9-point circle

Every triangle has a 9-point circle, which gets its name from the nine points this circle contains. Three of the points are the midpoints of the sides. The feet of the altitudes are three more points. The last trio comprises the points halfway between the vertices and the orthocenter. The easiest way to construct this triangle is to use the perpendicular bisector construction on each side, finding the midpoints. Then, we construct the circle through these three points. (By the Sword Theorem, all we need is the perpendicular bisectors of two segments joining midpoints.)

There are special cases of the 9-point circle. We can make the 9-point circle look like an incircle. We can get the 9-point circle to pass through one vertex of the triangle, but never two. Such considerations are a good source of test and homework problems. There are several proofs of the properties of the 9-point circle. We include a short proof using cyclic quadrilaterals here.

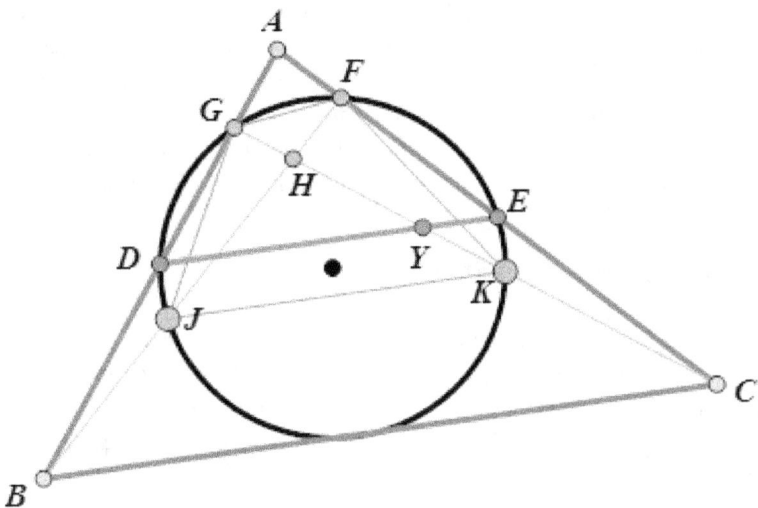

The 9 pt. circle for triangle ABC.

The figure above has been set up for the proof. We are given the outermost triangle, $\triangle ABC$. We are about to prove that, for any triangle, three trios of points all lie on one circle. These trios are the midpoints of sides (D and E are midpoints of sides), the feet of the altitudes (F and G) and the midpoints of the segments joining the vertices to the orthocenter H, the point where the altitudes intersect (K and J). The attentive reader will notice we did not complete any of these trios in the figure. Our plan starts with using the sides meeting at vertex A as any two sides of the triangle. We will first show that the two midpoints and two feet of altitudes on sides of $\angle A$ all lie on one circle. Since the two sides could have been any two sides, we will have shown that the first two trios all lie on one circle. Then we will use the two feet of altitudes with the last two points and get another cyclic quadrilateral. This implies the middle and last trios all lie on the same circle. Since three points determine a circle, all these points must lie on the same circle, the 9 point circle. Let's see the details.

Proof: Remember, the circle in the figure is not really there until we prove it is. So we can't use inscribed angles or tangents. We can use those right angles AFB and CGB with the vertical angles to get $\triangle FAB \sim \triangle GAC$. This gives us cyclic quadrilateral $BGFC$ because similar triangles in this position have outer vertices on a circle. We do not actually draw this circle. Since the segment joining two midpoints of a triangle is parallel to the third side, $\angle GDE \cong GBC$ and $\angle FED \cong \angle FCB$ because they are corresponding angles. Then quadrilateral $GDEF$ is also cyclic for the same reason as quadrilateral $BGFC$ and that circle is going to be our

9 point circle.

We urge the fatigued reader to stay tough. Yes, we have another quadrilateral to tackle. But, a quick look shows us that the situation we are about to handle is the same as the one we just did! Consider the triangle using vertex H, $\triangle BHC$. The points K and J are midpoints and F and G remain feet of altitudes for extended sides. This is the exact same given as we started with in the above paragraphs, for a different triangle. We will reach the same conclusion, that $JGFK$ is cyclic.   ■

1.11   Constructing specific lengths

Compass and straightedge can be used to construct specific sizes instead of geometrical objects. The three moves we will learn today start with one or two given sizes, a and b, along with a given unit length and end with segments with lengths ab, $\dfrac{a}{b}$ and $\sqrt{a}$. The reader should be able to prove all three of these constructions work.

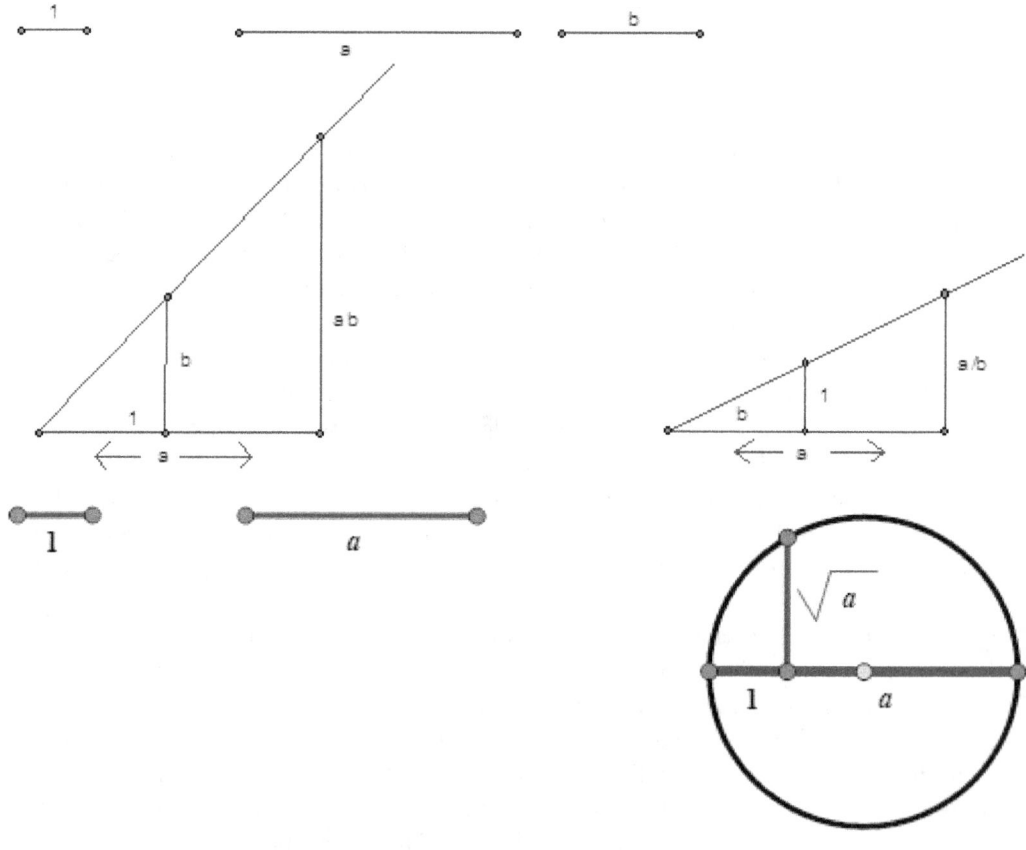

The reader is urged to try these constructions in *SketchPad*, making the illustrated lengths part of the drawing. It is a good exercise in conversions to use some arbitrary length for the unit and then use the calculator to adjust the measurements in terms of that unit. (Just divide any desired length by the unit length to get the desired length in terms of the given unit.)

Now that we can construct $\frac{a}{b}$, we could divide a given segment with length a into n equal pieces without too much trouble. There is, however, another construction for this same action and it is worth learning about. With the segment of length a drawn in a place with some room below it, we get to lay down our straightedge at any angle and draw a ray descending from the left-hand endpoint. Then we make our compass any length we want and mark out n equal segments along this ray. When we connect the last endpoint of these n segments with the right-hand endpoint of the given segment of length a, we have a triangle which has one side marked into n equal pieces.

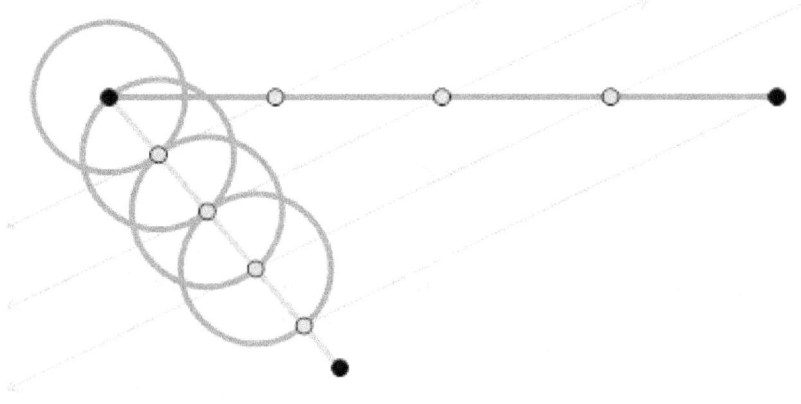

While the reader is thinking about the situation, let's take a moment to note a slight complication in the *SketchPad* version of this construction. In the illustration above, we have to cut the second segment into equal pieces. In this case, we had to chain four circles along the ray, using the point of intersection as the next center, and the previous point of intersection as the radius. Finishing the construction is the same using tools or SketchPad: we have to copy the lower left angle three times, once at each intersection point on the ray. Finishing in *SketchPad* is easy because we can use the construct parallel line command instead of copying angles.

Now that we have such nifty constructions, what can we make? We should be able to construct a length which is any composition of the arithmetic operations and square roots. Let's think about an example, say, $\sqrt{(3 + \sqrt{2})}$. We would first have to construct the length $\sqrt{2}$, which reminds us of trigonometry. Let's recall that the diagonal of a unit square has length $\sqrt{2}$. We could construct two sides of that square, or we could do the circle construction for square roots. Then we add 3 units to its length and perform the square root construction to take the square root of the whole thing. Here's one way it could have turned out.

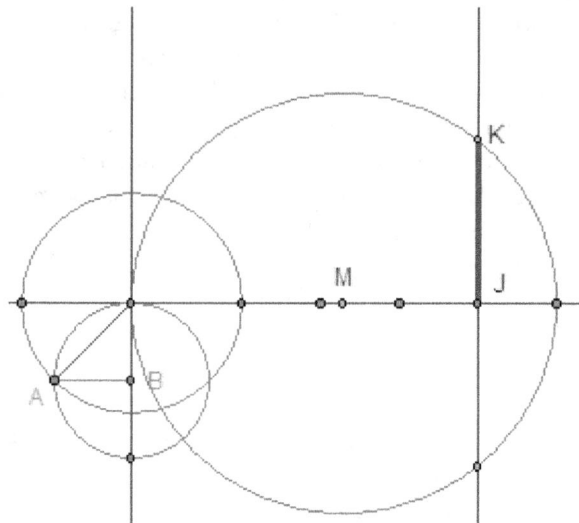

The segment $\overline{AB}$ stands for the unit segment. There is an isosceles right triangle whose hypotenuse has length $\sqrt{2}$. This hypotenuse has been used as the radius of a circle. One horizontal radius of this circle has been added to three units and then one more unit, whose left endpoint is J , has been added to the end. This sets us up for the square root construction. We found the midpoint of the horizontal segment whose length is $3 + \sqrt{2} + 1$ and labeled it M. The perpendicular at J meets this circle at K. The length of JK is the desired length, $\sqrt{(3 + \sqrt{2})}$.

Such constructions are like little puzzles where the only challenge is coming up with clever ways to put the pieces together. They beg the more important question: what sorts of things are beyond our ability to construct?

That turns out to be one of the questions which runs through mathematics from 300 BC to the present. The Greeks had some projects which defeated their geometric ingenuity. They could not trisect an angle with compass and straightedge. They could not construct the cube root of a given length. There were some regular polygons which they could not construct. They could not construct a circle and square with the same area. Plus, they did not know why such results were beyond their abilities.

Their mathematics was deep enough for them to combine ideas with the compass and straightedge in order to achieve the results. Some calculus courses include the Cissoid of Diocles, which aided the construction of a trisected angle. Investigations along these lines continue to this day. The French mathematician Lebesgue, famous for his generalization of the integral, thought the French results regarding such constructions so important that he made getting these ideas out of France a top priority as the Germans were advancing on Paris in World War II.

We will be focusing on compass and straightedge only. So we find the work of Gauss particularly useful. He was able to prove exactly which regular polygons were constructible and which were not, settling one ancient question. The number of sides has to be 2 to a power or 2 to a power times single powers of Fermat primes. The first five Fermat primes are 3, 5, 17, 257 and 65537. After he proved that the regular 17-gon was constructible, he found a construction for it. He was so happy with this result that he requested the construction be reproduced on his tombstone! This is not so surprising when we realize that his research linked him with great minds in the past and the future, resolving difficulties long thought to be mysteries.

His work also explained why the cube root construction was unattainable. Imprecisely, he was able to capture the compass and straightedge algebraically and thus show that odd roots are outside the reach of the tools. Galois and Abel contributed to this effort by completely solving the reducibility of polynomials questions, using abstract algebra.

Trisection of angles is worth a little more thought because each year, beginning geometry students recreate a construction which does not give trisection. It only **looks** as if the trick succeeds for acute angles. We will examine this little hassle in two pages.

Given angle ABC, can we construct two rays which cut the angle into three equal angles? The proof that the answer is "no" lies beyond *Geometry by Construction*. Also, an examination of methods which have been found for trisecting angles using, for example, a marked straightedge and a compass are not included because we have plenty to study in legal constructions themselves.

Learning a little bit more about the limitations is worth a few pages. Here is a brief summary of the most famous impossible constructions and their related topics.

(1) Squaring the circle. Given a circle, we cannot, in general, construct a square whose area is equal to that of the circle.
(2) Doubling the cube. Given a cube, construct a cube with twice the volume. This required the construction of a cube root. We have seen a square root construction. It turns out that composing the square roots describes all the roots we can construct.
(3) Constructing regular polygons. A regular polygon has n equal sides and equal angles. There is an easy calculation for finding the measure of one angle of a regular polygon but that does not usually help us construct one of those angles. We can construct any regular polygon with the number of sides described above. The regular 7-gon is the non-constructible regular polygon with the lowest number of sides.

(4) Angle trisection. Impossible with compass and straightedge. We will see a common mistake regarding this problem soon.

(5) Alhazen's Billiard Problem. Given two points A and B inside a circle, the construction of an inscribed isosceles triangle with A on one leg and B on the other leg cannot be constructed in general. Constructions exist, however, for A and B in special positions.

The proofs of all these claims require group theory. The references contain several texts which explain the details.

Here is the erroneous construction which rookie geometry students often think of by themselves. The angle A was given. The point A was used as the center of a circle so that the sides of the angle are the same length. The segment opposite angle A has been divided into three equal segments and the rays from A through these dividers do not look like trisectors. Of course, they aren't.

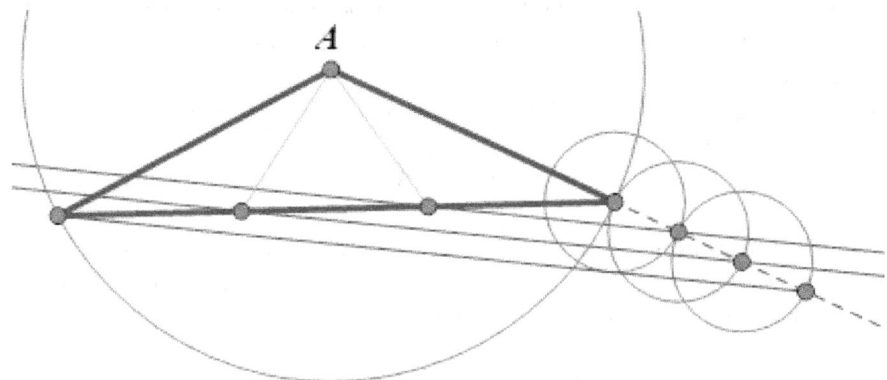

False angle trisection.

1.12   The Pythagorean Theorem

The Pythagorean theorem has so many proofs that there are books filled with them. We will look at two classics. First, we state the theorem. $\triangle ABC$ has a right angle at C if and only if the lengths of the sides satisfy the equation $a^2 + b^2 = c^2$.

Most people do not think of this theorem as an if and only if. We will look at a few proofs in the usual direction to start. The first proof takes four copies of the triangle and arranges them in a square. It is important to understand that such a configuration is indeed a square. The reader should know why each corner angle is 90 degrees and why each corresponding side is the same length.

The proof for this configuration is homework.

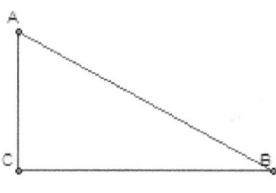

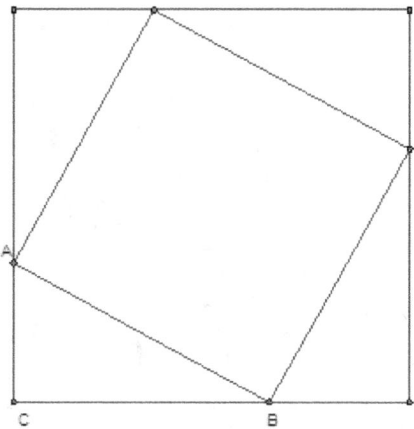

If the reader were to browse the hundreds of proofs of the Pythagorean theorem, he or she would notice many proofs use area cleverly. For instance, the strategy behind the figure above is to calculate the area of the large square two different ways. Since the area of the large square doesn't change, these calculations may be equated. When we just mull over the formula, using area makes sense because we have to get those squared terms somehow. Since a is the length of the side opposite $\angle A$, a^2 has a pictorial meaning which we will see in the figure below.

The second proof is from Euclid's Elements. The given triangle has a square constructed on each side of the triangle. The altitude from C has also been constructed. The proof, while longer than the one which goes with the above figure, has the benefit of showing how the largest square can be cut into areas which equal the areas of the smaller squares. And that is a huge hint concerning how to do the proof, in the homework set, of course. The reader will also need the usual formula for the area of a triangle.

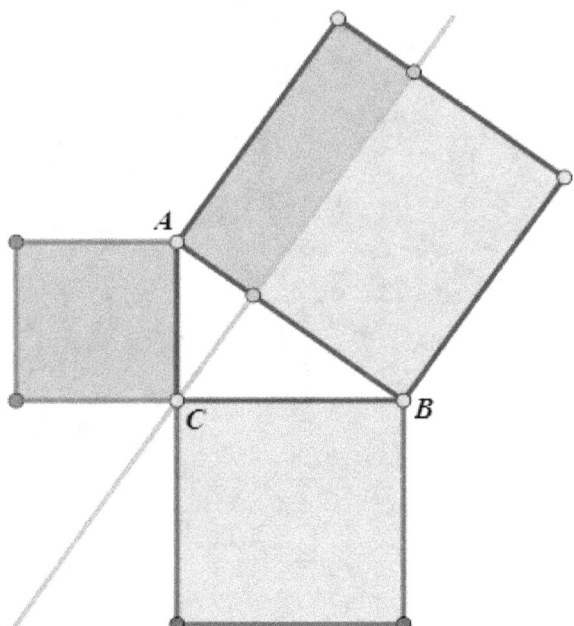

Euclid's proof of the Pythagorean theorem.

The Pythagorean theorem has thousands of related geometry problems. Since the Euclidean distance formula is the Pythagorean theorem, any math question involving measurement of lengths often needs it. For instance, recall the related rate, optimization, and hydrostatics sections in calculus used the theorem whenever the figure had a right triangle and lengths of sides were important. Virtually every SAT or GRE quantitative test has a few questions using this formula. Journals and magazines for math teachers have articles on it almost every month. Math historians still find old references to it as they dig through ancient documents.

Yet most people lack a solid understanding of the Pythagorean theorem. When the Scarecrow gets his diploma in The Wizard of Oz, he wildly misspeaks the theorem and most viewers do not even notice. There is much more to this gem of a theorem than squaring some sides. First of all, it is absolutely mandatory to know that the converse of the theorem holds. This is the "other direction" of the if and only if. Let's take a minute to prove that

if $\triangle ABC$ has side lengths a, b, c such that $a^2 + b^2 = c^2$, then the triangle is a right triangle.

Proof: If we take the sides with lengths a and b and make them the sides of a right angle, we can draw a right triangle whose sides are a and b but whose hypotenuse must be different from c. Let's call this new distance d. By the original Pythagorean theorem, $a^2 + b^2 = d^2$. But then $c^2 = d^2$ and we have our contradiction. The only way c and d can be different would be for one of them to be negative, which is not an option. Since our proof never used the obtuseness of the angle C, the angle could just as easily been acute and our proof would work the same. So we are done. ■

1.13 Circle Properties

Circles have central and inscribed angles. They have tangents and secants, chords, radii and diameters. It is essential to have a full understanding of all these terms and how they are related. Plus, there are some surprising relationships between these objects.

The Power of the Point might sound strangely familiar because of the commonly used software. It is the standard name for a property of points and circles. We start with a point outside a circle and any two secants or tangents containing this point. We find that there is a number calculated from a secant or tangent which remains the same for all others through this point.

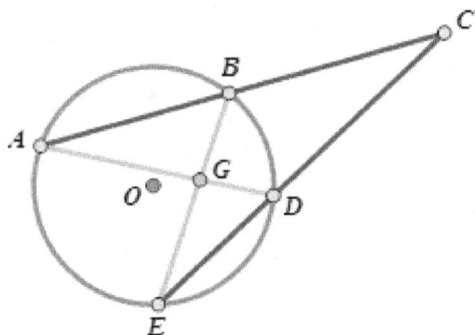

The power of the point.

We will prove the Power of the Point calculation does indeed work as the way it looks in the figure, but the proof is for homework. This drawing contains two extra segments which make the proof a breeze. The formula is $AC \times BC = EC \times DC$.

1.14 Inversion across a circle

Inversion across a circle is a move we can do with any circle. We will usually treat the circle as the fixed object, and we label the center O to emphasize its importance and make the circle easy to find. The definition is this

A point B inside or on a given circle has a unique point, B^{-1} or B' , collinear with B and with O, the center of the circle such that $OB \times OB^{-1} = r^2$, where r is the radius.

The construction for B^{-1} is an absolute must-know for success in our future work. This construction is a bridge between Euclidean, hyperbolic and elliptic geometries – the three big sections of this book. The successful reader will have to be able to do this construction in *SketchPad* and by compass and straightedge, without reference to notes. A thorough understanding of the construction leads to the quick solving of many future problems.

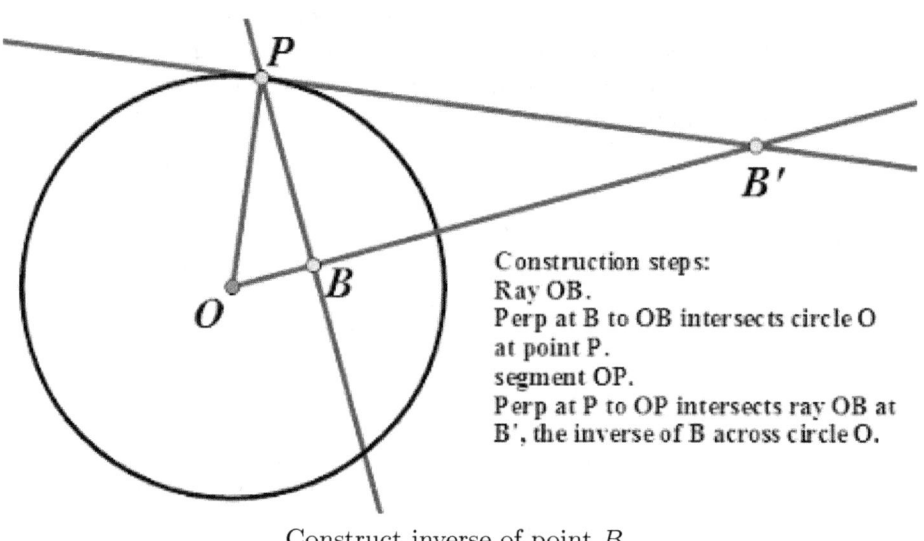

Construction steps:
Ray OB.
Perp at B to OB intersects circle O at point P.
segment OP.
Perp at P to OP intersects ray OB at B', the inverse of B across circle O.

Construct inverse of point B.

The proof that this construction indeed gives the inverse of B relies on a pair of similar triangles. (This should not be surprise, since the definition looks like the cross-multiplication of a proportion.) All we need is $\triangle OPB \sim \triangle OB^{-1}P$ and our proportion will be just right. The actual steps will turn up in the homework, of course.

We will soon prove two important properties about inversion which will combine to form an if and only if statement which we will use a lot in section 2. But first, we need to learn what orthogonal circles are.

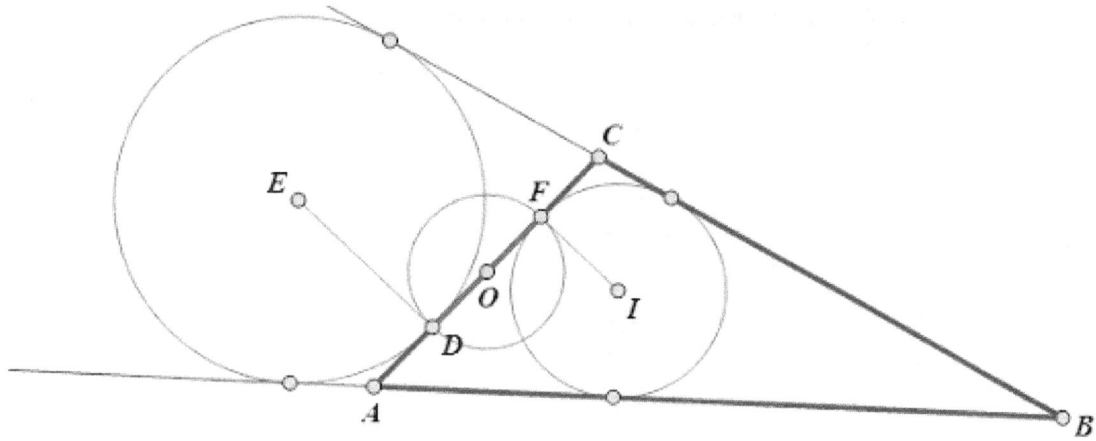

Fig. 1.1   A circle orthogonal to incircle and excircle.

Some readers will have seen orthogonal curves in calculus. Some readers may have explored a museum or gallery with curved walls where the curved walls meet at corners. These corners, when viewed close-up, might look like a right-angled corner. We all remember using calculus to find the slopes of tangents to curves. Orthogonal circles are circles whose radii are perpendicular at the points of intersection. In other words, a line tangent to one circle through a point of intersection contains a radius of the other circle, and vice-versa. The gallery walls may have been arcs of orthogonal circles; the corner was a point of intersection.

We will take the incircle and an excircle of a triangle as our example of orthogonal circles. The sharp reader will recall that these circles usually don't meet, so they cannot be orthogonal to each other and this is true. But, for scalene triangles, we can construct a circle which is orthogonal to both. We need an inradius and an exradius drawn to the side which is tangent to the given incircle and excircle. If we use the endpoints of these radii as the endpoints of a diameter, this new circle is orthogonal to the incircle and excircle. To verify this, we observe that the radius of the incircle is tangent to the new circle, and the same holds for the excircle.

We use the word orthogonal even though its meaning is essentially the same as the more familiar word, perpendicular, because perpendicular objects have always been straight, one-dimensional objects. We will continue this convention of using orthogonal for bent objects which meet at right angles. In fact, our bent objects will always be arcs of circles. We will now prove an important property about orthogonal circles.

Given a pair of orthogonal circles with a ray from the center of one circle acting as a secant to the other circle, the two points of intersection are a point and its inverse across the circle whose center starts the ray.

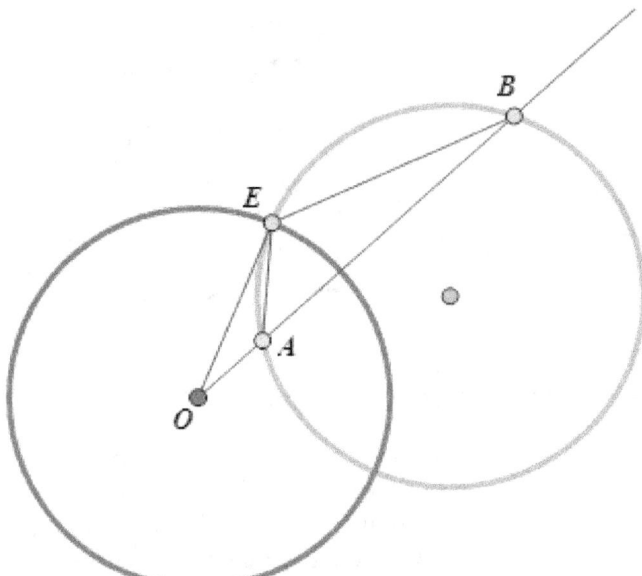

$\overrightarrow{OA}$ contains the inverse of A across circle O.

We have to prove what the calculation implies. Readers who worked through the previous page should have a good idea: we have to use point E. Since we are given orthogonal circles, we will have to use the point of intersection and at least one radius. Before moving on, we should think about strategy. Since we need an equation using products of lengths, we should look for similar triangles. A proportion of sides might turn into the equation we need.

Proof: Remember, we need triangles whose sides include $\overline{OA}$, $\overline{OB}$ and $\overline{OE}$ twice so that we can use the definition of a point and its inverse Certainly, we can have angle O in common to two triangles, $\triangle OEA$ and $\triangle OBE$. An attempt was made to name these triangles in their corresponding order. Angle OEA and angle EBA both cut the same arc, the arc AE. Both these angles are inscribed in the right-side circle. So these two angles are the same size! Since we matched E in the first triangle with B in the second, our correspondence is correct and the triangles are similar in the order they are named.

$$\frac{OA}{OE} = \frac{OE}{OB}.$$

Cross-multiplication shows us that the point B is playing the part of A^{-1} in the definition of the inverse of a point. ∎

This proof tells us that any line through the center of one pair of orthogonal circles creates a point and its inverse. This will be most useful in section 2. Now we will see that the situation works in reverse:

If a circle passes through a point and its inverse across circle O, then the circle is orthogonal to circle O.

Proof: This time, we are starting with two circles, one of which passes through a point and its inversion across the other circle. We seek to prove that circle C is orthogonal to circle O. This is the converse of the theorem just proved on the previous page. The drawing is not exactly the same: we have replaced the label B with the label A^{-1} because that is what the point is. People with mathematical experience should think that we might be able to back-track through the last proof and this is a good instinct.

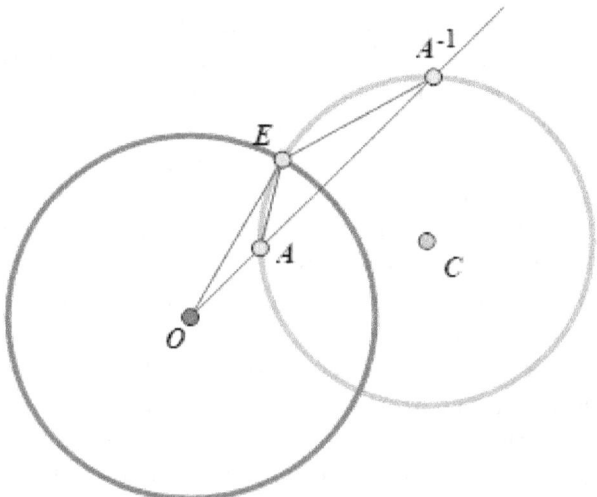

The equation $\dfrac{OA}{OE} = \dfrac{OE}{OA^{-1}}$ is the given information, this time. (We ended with this information last time.) Since we derived this equation from similar triangles last time, we look for similar triangles to happen again. And they do, but not for the same reason. Take a close look at the positions of these segments. Each pair forms the sides of angle O in each of the triangles we used before. This means we have the sAs version of similar triangles, which we last saw as a consequence of the theorem of Menelaus.

$\triangle OEA \sim \triangle OA^{-1}E$ implies that $\angle OEA = \angle OA^{-1}E$. Since $\angle OA^{-1}E$ is inscribed in circle C, it is equal to half its arc. Then $\angle OEA$ is also equal to half its arc. Therefore, $\angle OEA$ is also an inscribed angle, which forces $\overline{OE}$ to be tangent to circle C. Since the radius is tangent, the circles must be orthogonal.   ■

These two proofs illustrate the strong relationship between inverse pairs of points. Now we have the powerful fact that a circle containing a point and its inversion across a circle is orthogonal to the circle of inversion. This will be most useful in when we need to prove two circles are orthogonal. The construction itself is from Goodman-Strauss's article.

1.15   Miquel circles and Miquel point

We now have enough vocabulary to explore geometry beyond the usual secondary course. The trip we are about to start through the ideas of Auguste Miquel resembles many other digressions awaiting attention. The journey begins with a theorem we could run into at a math conference, in a journal, on a webpage, or in a reference book. It catches our interest enough to wonder why it might be true. The proof is not so bad and we wonder if there's more.

That's one thing nobody worries about in mathematics: of course there's more. Not only did Miquel have other theorems, but we will find a modification of one theorem which fits our orthogonal circles in an unexpected way in the next section. And that is one of the ways mathematics happens. Our journey begins with the Miquel point, a point where three circles related to a triangle meet. Since the Miquel point appears in some college geometry, it is not obscure knowledge.

Choose one point on each side of $\triangle ABC$. Label these three points D, E, F. Three non-collinear points determine a circle. So we construct the circle through each vertex and the points on the sides adjacent to that vertex. These three circles meet in the Miquel point, Q. A drawing will help a lot in this case. In fact, we have the construction of the Miquel point in three steps.

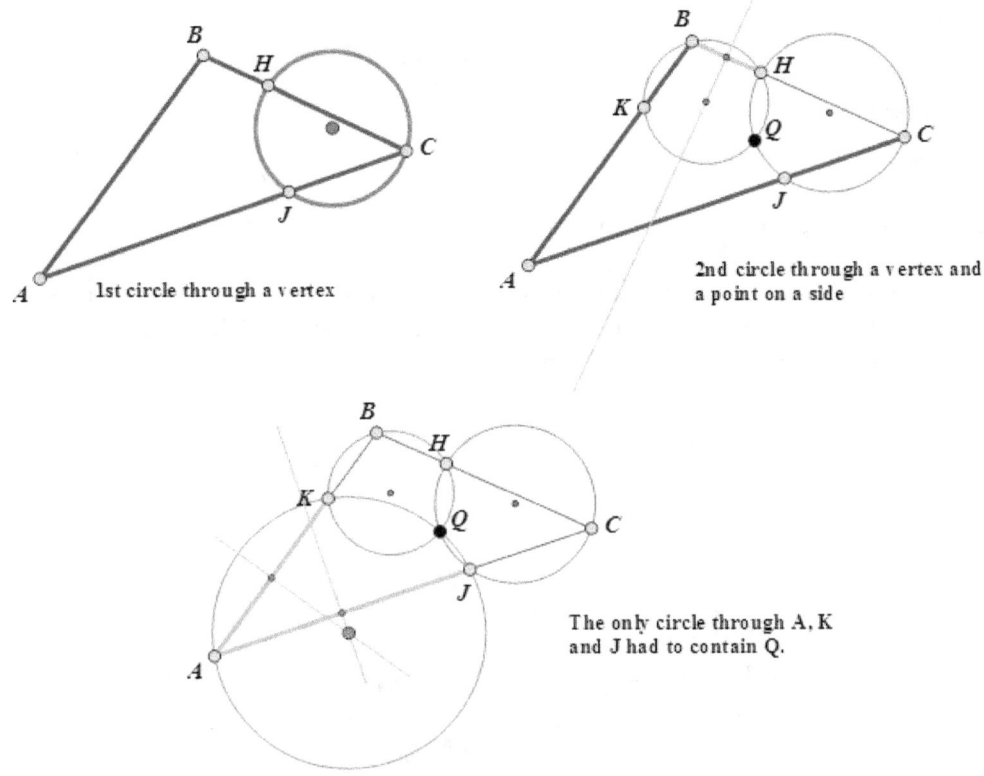

1st circle through a vertex

2nd circle through a vertex and a point on a side

The only circle through A, K and J had to contain Q.

Miquel point construction.

Proof: The above figure has three drawings in order to show the steps. The points H, J, and K could be anywhere on their respective sides; more precisely, the placement of the first two circles in the construction decides everything and it does not matter which two circles are found first. The first two circles in the figure use the three points H, J, K and two vertices. Each of these circles is determined by three points. These first two circles must have the point H in common, as well as a second point, which has to be Q. To finish the proof, all we have to do is show a circle goes through the unused vertex A, its two neighboring points J, K and the Miquel point Q. Well, that should sound familiar. We seek to prove that quadrilateral $AKQJ$ is cyclic.

So we make it our goal to get $\angle A$ supplementary to $\angle KQJ$. Since the other two quadrilaterals which look like the one we are working on are cyclic, we have $\angle C$ supplementary to $\angle HQJ$ and $\angle B$ supplementary to $\angle KQH$. Substituting this information into the sum of the angles A, B, and C is 180 degrees gives us exactly what we need. (The reader should do the algebra.)　　■

1.16　Undergraduate research result

Here's a nice problem from student Megan Ternes which combines inversion across a circle with Descartes's Theorem for mutually tangent circles. The theorem says if four circles are mutually tangent then the curvatures k_1, k_2, k_3, k_4 satisfy the equation

$$(k_1 + k_2 + k_3 + k_4)^2 = 2\left(k_1^2 + k_2^2 + k_3^2 + k_4^2\right) \tag{1}$$

Any reader handy with constructions is urged to attempt to construct four mutually tangent circles: some planning is required. All the construction marks for our example are in the figure below: a remarkably economic use of a few circles and lines in order to get four circles, each one of which is tangent to the other three. These are circles centered at A, B, G and C. (There are other ways to arrange four mutually tangent circles besides one circle containing the other three.)

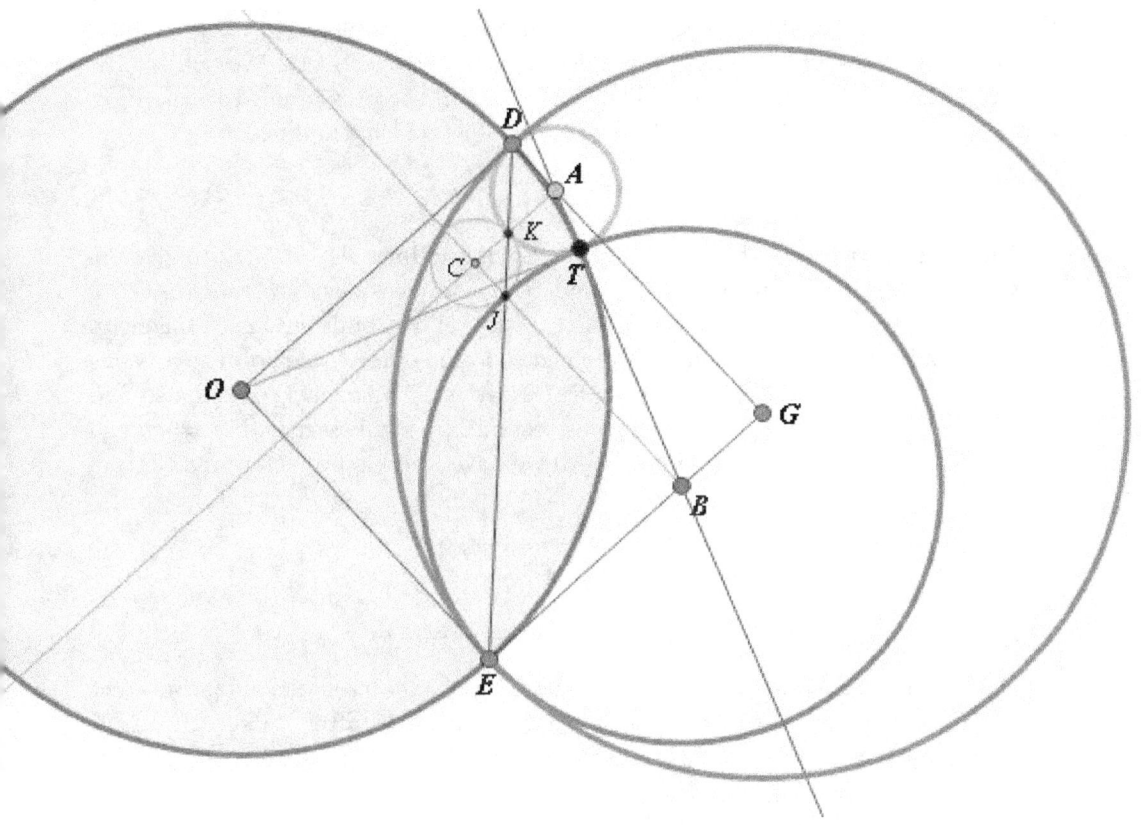

Descartes's theorem example.

Megan used the orthogonal circle structure to prove circle C is indeed tangent to circles A and B at the points on the radical axis $\overline{DE}$ of circles G and O. The construction starts with circle O and a pair of perpendicular radii. Finishing the square gives us G, the center of a circle orthogonal to circle O and with the same size Euclidean radius as circle O. The position of point A can be anywhere along that side of the square and the construction holds. The choice of A decides the tangent at T used to find point B, giving two circles A and B orthogonal to circle O.

For this arrangement of circles, the radical axis of circle O and the circle G, the same size as circle O, intersects these two interior circles at the points of tangency of a fourth circle, which is also tangent to the circle congruent to circle O.

We now verify Descartes's Theorem for this figure. The theorem says that the curvatures of four mutually tangent circles satisfy the equation (1). There is one more little hassle: when one of the mutual tangent circles contains the other three, its curvature takes a negative sign in the equation. Some authors call this signed curvature the *bend*. Our example has one outermost circle G containing the other three circles; so we must use a signed curvature for that circle.

Let's assign the radii a and b to circles A and B, respectively. We will set the radius of the two big circles to 1. The Pythagorean Theorem applied to the right triangle ABC gives us $(a+b)^2 = (1-a)^2 + (1-b)^2$ which simplifies to

$$ab = 1 - a - b \tag{2}$$

Our construction uses the meeting of the Euclidean lines $\overleftrightarrow{AK}$ and $\overleftrightarrow{BJ}$ to give the center C of a circle tangent to circles A and B. The dimensions of rectangle $AGBC$ show the radius of the fourth circle is also $1 - a - b$! We built circle C tangent to circles A and B; we will now verify it is tangent to the outer circle because it is one of exactly two sizes available, according to Descartes's Theorem. (We will also find the circle of the other available size in this section.) We will start with $r_3 = ab$ and the bend of the containing circle is -1. All we have to do is prove formula (3).

$$\left(\frac{1}{a} + \frac{1}{b} + \frac{1}{ab} - 1 \right)^2 = 2 \left(\frac{1}{a^2} + \frac{1}{b^2} + \frac{1}{a^2b^2} + 1 \right) \tag{3}$$

On the left, we add the fractions. When we substitute (2), we get $\left(\dfrac{(2a+2b)}{ab} \right)^2$.

On the right-hand side of (3), adding the fractions matches denominators right away. The numerator turns into the left-hand numerator after we substitute (2) into the $(ab)^2$ term. Equation (2) describes the relationship between the radii of two of the circles and happens to be the radius of a third circle. Using radius 1 for the fourth circle satisfies Descartes's Theorem.

There is another circle we can construct tangent to three circles in this figure, with very little work: the inverse of circle C across circle O. Since inversion preserves incidence, there are three points ready to be used with the given information to find three points on the inversion of circle C. The eager reader will find this construction in the homework.

1.17   A cyclic example

That undergraduate research problem does not resemble the homework. Homework problems are meant to get done on one sheet of paper. They are also built to elude the internet support by being too specific or obscure for anyone to bother writing about. For success, keep the Rules and chapter ideas handy. Always take the time to understand the given. Constructing an example is not a waste of time; however, an example does not prove the general case of anything.

Our first example relies on early concepts from this section. Given cyclic quadrilateral $ABCD$ with point E on extended side $\overline{DC}$ as pictured, prove angle $\angle BCE = \angle BAD$.

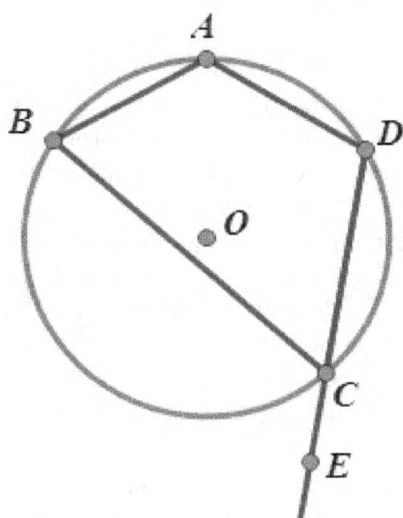

We know that opposite angles of a cyclic quadrilateral are supplementary. Just that vocabulary word should make a connection with a structure in the drawing: angles BCE and BCD are a linear pair and thus supplementary. We are supposed to get $\angle BCE$ the same size as $\angle BAD$ and we just found a connection between them: $\angle BCD$. Let's write out the thinking and see if we have enough to be finished. So, here's what a person would actually write for an answer.

> Angles BAD and BCD are supplementary because they are opposite angles in a cyclic quadrilateral. Angles BCE and BCD are supplementary because they form a linear pair. Angle BCE equals angle BAD because they are supplementary to the same angle.

Imprecise writers often fumble problems like this when they call angles supplementary because they **are** supplementary. Let's take a look at an illogical answer and derive a warning against fuzzy thinking.

> Angles BAD and BCD and BCD and BCE are supplementary. So they all add up to 180. We can subtract BCD from both and that leaves BCE = BAD.

This is one of those answers which turns a professor's hair grey because the ideas in the sentences are correct but the language is not clear. The work is not totally wrong; but neither is it good enough to be correct.

The first sentence tries to list two pairs of supplementary angles but does not mark them off by pairs. Also, the first sentence does not tell why the angles are supplementary. Sure, the pairs are supplementary; but the writer has the burden of telling why. The book's author would like to blame the second sentence on social media which uses pronouns far too often. The second sentence is wrong because all four angles actually add up to 360, which we do not need at all. The third sentence refers to an unwritten equation, probably on the writer's scrap paper. In fact, $\angle BAD + \angle BCD = 180 = \angle BCD + \angle BCE$ is a perfectly true equation, once we know why the angles are supplementary. Subtracting $\angle BCD$ from both sides leaves the desired result, so the conclusion has been reached.

We gain three pieces of advice: use nouns, use course vocabulary and support the statements.

Another cyclic example

With all the cyclic ideas fresh, let's try another problem. Prove that if cyclic quadrilateral $JGHI$ has diameter $\overline{JG}$, then angles H and I must be obtuse.

We can imagine a picture, sketch one or construct one. If we see the situation, we can answer directly. With or without a picture, is the next answer good enough?

Angles H and I are inscribed in a circle and each cuts more than a semicircle. So each is more than 90 degrees.

Right away, we have a strong sentence which pretty much finishes the problem. The writer's instinct to explain should have triggered: we need a "because" after the word "semicircle." We can say that, once we have that reason, we would be done. Try this correction.

Angles H and I are inscribed in a circle and each cuts more than a semicircle. Consider angle H, also known as angle GHI. The point J must be between I and G because of how we name the quadrilateral. Thus, angle H cuts more than a semicircle, as does angle I. So each is more than 90 degrees.

A picture would clinch the argument; but all we need is there. That "consider" is a very handy word for mathematicians who want to write something complicated about, in this case, angle H. Another way to capture this meaning would be, "focus on angle H."

1.18   An example from orthogonal circles

Let circle O intersect circle P at points A and B. If circle O is orthogonal to circle P at A, prove the circles are orthogonal at point B as well.

   We have been assuming this property all along because we have had numerous orthogonal circles and we have always treated them as orthogonal at both places of intersection. Now, we need to recall what makes circles orthogonal: their radii are perpendicular at the point where they intersect. Apparently, this property applies to both points of intersection; but that is what we're trying to prove. We need a picture which captures the given.

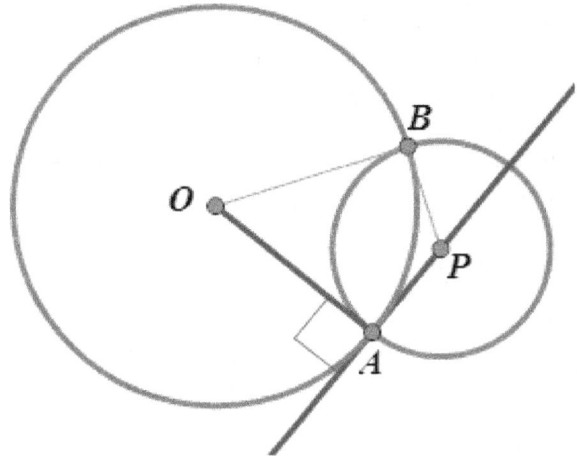

   The reader might want to write that $\angle OBP$ must be 90 degrees because the radius drawn to the point of tangency is perpendicular to the tangent. That would be circular reasoning, however, because we don't know segment $\overline{BP}$ is tangent to circle O at B. We know the two radii are perpendicular at A, but not at B. We have to pretend we don't know $\angle OBP$ is 90 but find a reason based on the given to force $\angle OBP$ to be 90.

   We have $OA = OB$ and $PB = PA$ because they are radii of the same circles. We have the given 90 degree angle. Perhaps congruent triangles comes to mind. All we need is $\overline{PO}$ drawn in. Let's go for the full answer.

   We have radii of the same circle $OB = OA$ and $PB = PA$.
   Two points determine a segment, so we construct
   $\overline{OP}$. $OP = OP$ by the Reflexive Property of Equality. Then
   triangle POB is congruent to triangle POA by SSS. Angle PAO
   equals angle PBO because they are corresponding parts of
   congruent triangles. Since angle PAO was given 90, angle PBO
   must be 90 degrees as well.

1.19   Summary

All terms and theorems in the Geometry Preliminaries pages.

Construct the tangents to a given circle from a point outside the circle.

Definition and properties of a cyclic quadrilateral.

Uses of similar triangles, AAA, sAs.

Menelaus and Ceva's theorems.

Incircle, excircle, circumcircle, 9-pt circle.

Orthocenter, circumcenter, centroid, incenter and excenter constructions.

Construct the lengths $\frac{a}{b}$, ab and $\sqrt{a}$, given lengths a, b, and 1. And construct compositions of these functions.

Divide a given segment into n equal pieces.

Impossibility of angle trisection, squaring a circle, constructing cube root.

Pythagorean Theorem and converse.

Power of the point.

Definition of inversion and relation to orthogonal circles.

Constructions of orthogonal circles.

Miquel point

Descartes's Theorem for tangent circles

1.20   Euclidean Geometry Homework

For constructions, show your construction marks. For proofs, use nouns and write only true things. Calculator, compass and straightedge required.

(1) Prove Trig Ceva: the three Cevians pictured are concurrent if and only if $\sin \alpha_1 \sin \beta_1 \sin \gamma_1 = \sin \alpha_2 \sin \beta_2 \sin \gamma_2$.

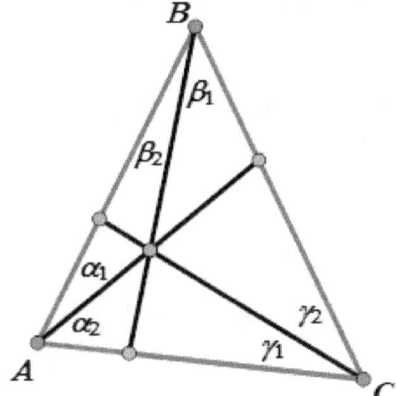

(2) Given right triangle ABC, construct three copies to form a square as pictured and prove your construction works. Use the figure to prove $a^2 + b^2 = c^2$.

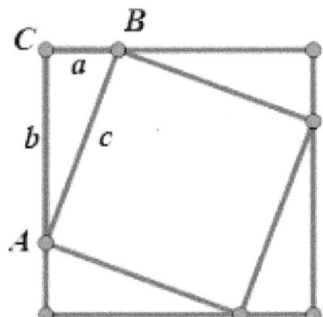

(3) Prepare to prove $a^2 + b^2 = c^2$ using the right triangle ABC as pictured. This time, begin the proof by showing that the extended altitude from C cuts the square of the hypotenuse into two rectangles with areas a^2 and b^2.

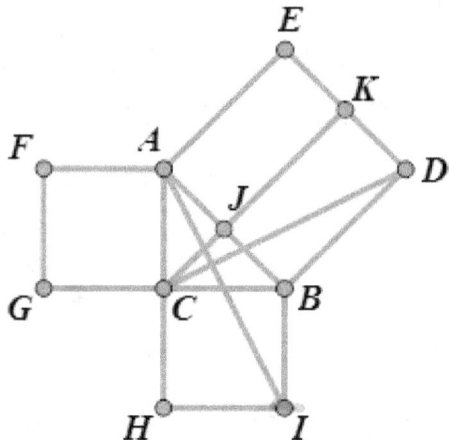

(4) Use the figure above and your work from problem **3** to finish the proof that
$a^2 + b^2 = c^2$.

(5) Prove the Power of the Point theorem.

(6) Prove that the formula for Menelaus's theorem holds when the line through D, E and F is perpendicular to a side of the triangle ABC.

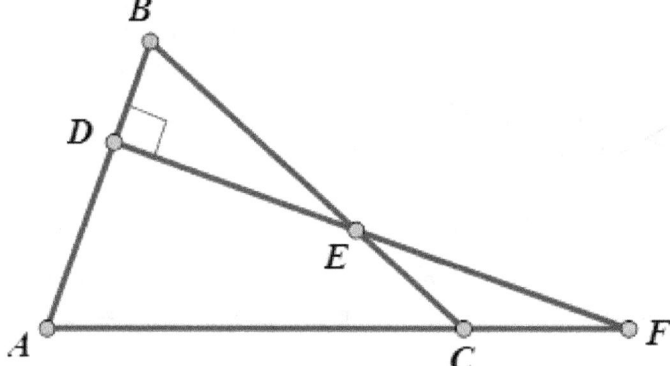

(7) Given isosceles right triangle ABC in the figure below, with the square of each side constructed, prove that the outermost six points all lie on one circle.

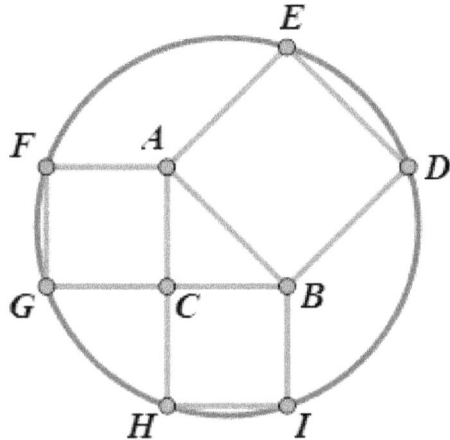

(8) Construct the figure for the problem **7**.
(9) Given quadrilateral $ABCD$ with opposite angles congruent, prove $ABCD$ is a parallelogram.
(10) Given quadrilateral $ABCD$ with consecutive angles supplementary, prove $ABCD$ is a parallelogram.
(11) Given quadrilateral $ABCD$ with diagonals which bisect each other, prove $ABCD$ is a parallelogram.
(12) Given quadrilateral $ABCD$ with opposite sides congruent, prove $ABCD$ is a parallelogram.
(13) Given quadrilateral $ABCD$ with one pair of sides congruent and parallel, prove $ABCD$ is a parallelogram.
(14) Each of $\mathbf{9} - \mathbf{13}$ has a true converse (switch the given and conclusion statements). Try those five problems.
(15) Given isosceles triangle ABC with base $\overline{BC}$, prove the altitude from A is also an angle bisector.
(16) Given isosceles triangle ABC with base $\overline{BC}$, prove the median from A is also an angle bisector and an altitude.
(17) The triangle ABC has midpoints H, G, F. Also, D is the midpoint of $\overline{BG}$. The points A, I and E are collinear, as pictured. Show that E is the midpoint of $\overline{BH}$.

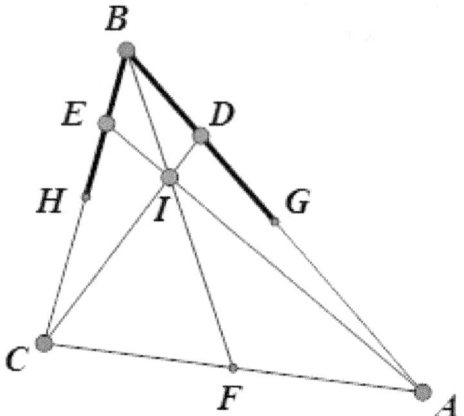

(18) Given trapezoid $ABCD$ inscribed in circle O with $\overline{BC}$ parallel to $\overline{AD}$, prove $\angle A \cong \angle D$.

(19) Given triangle ABC congruent to triangle DEF and $\overline{AB}$ is parallel to $\overline{DF}$, (as if a reflection occurred) prove the triangle AGE is isosceles.

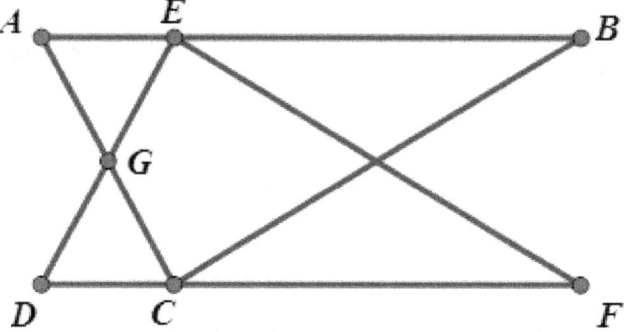

(20) From the definition of a point inverted across a circle O, recall that the points O, A and A^{-1} are collinear. Suppose point P is on circle O. Prove that $P^{-1} = P$.

(21) Suppose triangle ABC is isosceles with base $\overline{BC}$. Then its 9-point circle will have special position relative to the base. What is this position?

(22) A rectangular box is 8cm by 15cm by 144 cm. What is the longest stick which will fit inside the box?

(23) Suppose we want a circle with twice the area of a given circle with radius 5cm. What is the radius of the larger circle?

(24) Construct an equilateral triangle and then the square of each side. Then prove the outermost six vertices lie on one circle. (See problem **7**.)

(25) This time the figure has misleading structure. Do not be fooled! Follow the facts. Given congruent circles A and B with tangents $\overline{AC}$ and $\overline{DB}$, prove $ABCD$ is a rectangle.

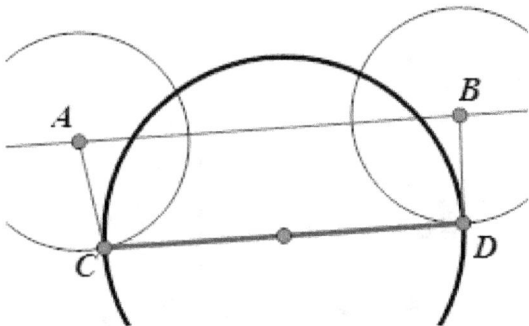

(26) Circle O has radius 3. Solve for x.

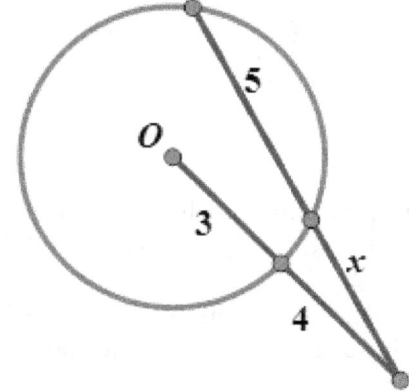

(27) The isosceles triangle ABC with base $\overline{AB}$ has Cevians $\overline{AD}$ and $\overline{BE}$ which cut the legs of the isosceles triangle at the same distance from the base and these Cevians intersect at F. The ray $\overrightarrow{CF}$ intersects the base at a point called M. Prove that point M has to be the midpoint of the base.

(28) Construct the circle tangent to circles A, G, and B in the Descartes' theorem figure.

(29) Construct an example of the Miquel point where the point Q ends up outside the triangle. Hint: don't choose all points close to midpoints for your points on the sides.

Here's an example of your goal.

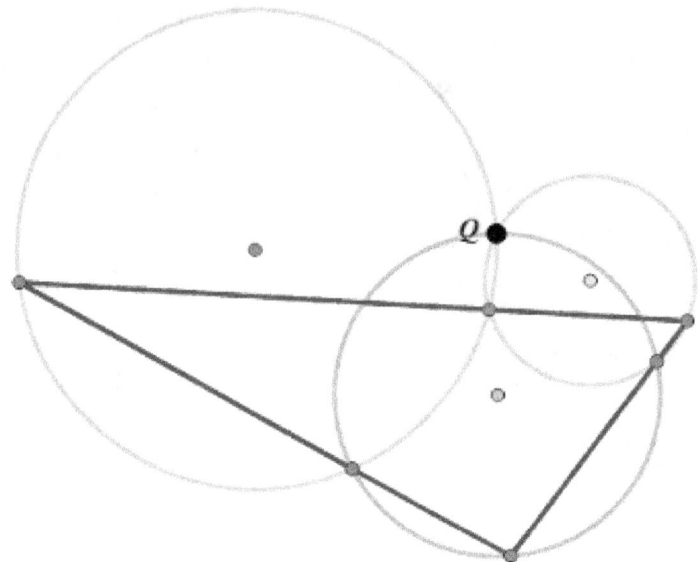

(30) Lindemann proved that π is transcendental in 1882, which implied that squaring the circle is impossible in Euclidean geometry. Exactly how are the constructible numbers related to the transcendental numbers?

(31) Prove Ptolemy's theorem: the product of the lengths of the diagonals of a cyclic quadrilateral is equal to the sum of the products of the lengths of the opposite sides.

(32) Use Pythagorean Triples to find distinct integers a, b, c, d, e (all greater than 1) which satisfy the equation $a^2 + b^2 + c^2 + d^2 = e^2$.

(33) How many solutions exist using positive integers a, b, c which satisfy the equation $a^3 + b^3 = c^3$. How do you know?

(34) Position three points, one on each side of a triangle (you choose the triangle) so that the Miquel point is the incenter of the triangle. This can be done by construction.

(35) Consider a triangle with its incenter and inradii drawn. How is a vertex angle of a triangle related to the opposite angle with vertex I?

(36) For a triangle ABC with Miquel point Q constructed using the three circles we see in the book, prove that the triangle with the centers of these three circles as vertices is similar to triangle ABC.

Chapter 2

Hyperbolic Geometry

Here's where we study our first bent space. Eighty years before Einstein's theories were confirmed at that solar eclipse, a few mathematicians were exploring options in geometry which had not been considered as options before. The options depended on the Parallel Postulate: through a point not on a given line, there is exactly one line parallel to the given line.

The tidy structure in Euclid's Elements motivated mathematicians to attempt to prove the Parallel Postulate. They desired the Elements to have one less assumption. The most fruitful strategy turned out to be proof by contradiction which then changed into the development of non-Euclidean geometry. In other words, proof by contradiction did not work as intended and, instead, opened a door into unexpected possibilities.

The Parallel Postulate has two negations because there are two options in its negations. In this section, we will use the following negation: Through a point not on a given line, there are many lines parallel to the given line. Saccheri, Lobachevsky, and Bolyai took this statement as true and then reasoned what else would be true. Instead of getting a contradiction, they found logical consequences. Lobachevsky and Bolyai discovered many facts about hyperbolic geometry but they did not have an example of a bent space where people could see pictures of hyperbolic entities. About 1900, Klein and Poincaré produced models of hyperbolic geometry. We will be using the Poincaré disk model.

Volumes have been written about the development of non-Euclidean geometries. The work certainly ranks amongst the highest intellectual achievements, and not just because it took a millennium to figure out. The fight against hyperbolic geometry flavored the philosophy and literature of the time. But once Klein and Poincaré gave us hyperbolic worlds to play with, the resistance crumbled.

Hyperbolic geometry is another example of mathematicians thinking ahead of the game. When that solar eclipse demonstrated that the universe wasn't Euclidean, mathematicians had a few examples of bent space ready to go.

The mathematicians of 1900 saw what we will see: geometry as an axiomatic system. (For the details of axiomatic systems, check out the Appendix.) We skip all the turmoil, controversy and struggle and go right to the good stuff. An axiomatic system is a set of axioms assumed to be true, with some key words designated as undefined terms. All other words take on their usual meanings. There can be further definitions, as well. The rules of logic apply, leading to theorems. Viewed in this way, Euclidean geometry and hyperbolic geometry differ by the parallel postulate, or fifth axiom. If an interpretation of the undefined terms allows a model in which all axioms are true, then we know the axioms do not contradict each other. The disk model was significant as a model of an axiomatic system because it used the same starting set of axioms as Euclidean geometry, except for the negation of the parallel postulate. This meant that the parallel postulate is independent of the other axioms – it cannot be proved or disproved from those axioms. Thus the tumult over the parallel postulate was resolved, but in an unexpected way.

The ramifications of the hyperbolic disk model influenced mathematicians to seek the logical foundations for all math. Sets of axioms built for this development were proposed, debated, and some were shot down in flames. The logical foundations of mathematics continues to be an area of active research.

The development of hyperbolic geometry was an important event in the history of mathematics and philosophy. Section 2 of this course explores the disk model, by construction. This means that we will always be able to see what we are talking about, a perspective which Bolyai and Lobachevsky lacked. The disk model exists inside Euclidean geometry – we will use Euclidean constructions to make hyperbolic objects. Here is a summary of where we begin.

2.1 The Rules of the Hyperbolic Disk

(1) All the rules of Euclidean geometry apply in all parts of all pictures.
(2) Hyperbolic Space consists of the points inside the circle O.
(3) Hyperbolic points are just like Euclidean points.
(4) There are two kinds of hyperbolic lines: diameters of O and arcs of circles orthogonal to O.
(5) If a hyperbolic angle has a curved side or sides, use tangents to measure angles.

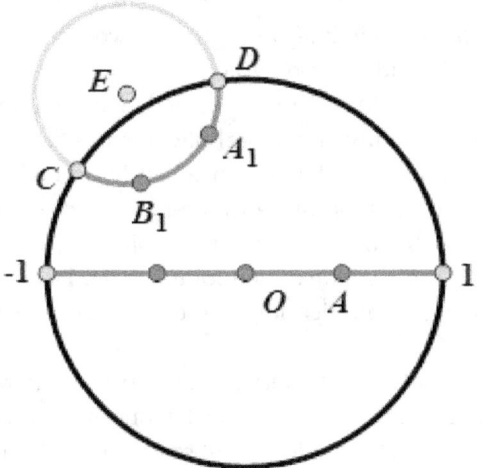

Examples of hyperbolic lines.

The illustration above emphasizes some crucial ideas. The hyperbolic points are inside the disk, including the points which make up hyperbolic lines $\overleftrightarrow{AB}$ and $\overleftrightarrow{CD}$. The boundary and points outside the boundary are not hyperbolic points. Point E

is the Euclidean center of the Euclidean circle which contains the hyperbolic line $\overleftrightarrow{CD}$. We need point E in order to draw the hyperbolic line; yet E is not a hyperbolic point. We will usually abbreviate these ideas by saying, "Hyperbolic line $\overleftrightarrow{CD}$ has center E." We also have the strange option of naming hyperbolic entities by using non-hyperbolic points: hyperbolic line $\overleftrightarrow{AB}$ and hyperbolic ray $\overrightarrow{OB}$ use points outside the disk in their names.

We should take a few minutes to verify that the first few Euclidean axioms are true in the disk model. We have a mental adjustment to make, first. From now on, we will be discussing hyperbolic objects. These hyperbolic objects are also Euclidean objects, since the disk is inside the Euclidean plane. For instance, we have hyperbolic lines and we have Euclidean lines. So we should use the word, "line" with its appropriate adjective, unless the context is clear. This means we will have some clumsy sentences to read and to write.

The first axiom, that two points determines a line, requires a few cases. If the two distinct points are on a diameter, in other words, they are Euclidean collinear with O, then the hyperbolic line through them is the diameter of circle O which contains them, minus its endpoints. If the two points are not on a diameter, then we construct the inverse of one of these points and then construct the Euclidean circle determined by these three points. Since the circle goes through a point and its inverse, the circle must be orthogonal to circle O, which means the arc of the circle inside the disk is a genuine hyperbolic line. The reader should think about why it does not matter which point we invert across circle O.

The construction of a hyperbolic line through any two points uses the inversion we studied near the end of Euclidean geometry.

The second axiom says that lines can be extended indefinitely. At first, this might look like a difficulty. Certainly any hyperbolic line has finite Euclidean length. But, as the reader might expect, we have to use the hyperbolic distance formula to measure lengths in the disk. It is quite different from the Euclidean distance formula because, to find the distance between two points A_1 and B_1, we need the coordinates of them and of the Euclidean endpoints, C and D, of their hyperbolic line. And if that is not enough complication, these coordinates are supposed to be complex numbers.

Fortunately, we can translate any hyperbolic line to the line which corresponds to the real (horizontal) axis. That way we can let one line stand for all and yet avoid the imaginary parts of the complex numbers. We can even send point B_1 to point O, so we can always measure from the origin. We won't even print the general hyperbolic distance formula because we won't need it. Here is the modified hyperbolic distance formula we will use:

$$OA = \left| \ln\left(\frac{a+1}{a-1} \div \frac{0+1}{0-1}\right) \right| = \ln\frac{1+a}{1-a}.$$

(We use a for the real coordinate of point A.)

People who do not judge by appearances have an advantage in hyperbolic geometry. For instance, hyperbolic segments of equal Euclidean length do not necessarily have the same hyperbolic length. In fact, it easy to construct a hyperbolic segment which is as small as a punctuation mark with hyperbolic length as large as we want. Another strange attribute of the distance formula is its reliance on points which, strictly speaking, are not in our hyperbolic space. Remember, the boundary circle is not included in our space. We will see that points on the boundary are so useful that we grant them a sort of honorary membership status.

Here is how to use the hyperbolic distance formula. Suppose the disk has radius 1 and $a = \frac{1}{2}$. We can calculate the length of $OA = \ln \frac{\frac{3}{2}}{\frac{1}{2}} = \ln 3$.

We leave the verification that hyperbolic lines can be extended indefinitely to the homework.

The hyperbolic distance formula implies that hyperbolic segments of equal-looking length can be of very different hyperbolic length because the closer an endpoint gets to the boundary, the longer the segment will measure. This is part of the idea behind many of the Escher circular tilings – the figures get smaller as they approach the boundary, but the figures are congruent in the hyperbolic sense.

The third axiom says that our geometry has circles, the set of points equidistant from a given point. Hyperbolic circles have to lie inside the disk, of course. They are the same as Euclidean circles in appearance. But the hyperbolic center of a hyperbolic circle is different from its Euclidean center, unless the center is O, in which case they are the same point. This kind of situation makes sense with the hyperbolic distance formula: the hyperbolic center will look closer to the boundary than the Euclidean center because the radius pointing directly to the boundary will have to look shorter than the radius pointing directly at O. Then they can have the same hyperbolic length.

Drawing a hyperbolic circle is easy: we simply put the point of the compass inside the disk, adjust the radius to fit inside the disk and draw the circle. The situation gets interesting when we try to find the hyperbolic center of this circle. This is the first of many opportunities to explore the hyperbolic disk using compass and straightedge. How are we going to find the hyperbolic center of the circle?

We will need some properties of circles which have to hold, even in hyperbolic space. For instance, all the diameters of a circle meet at its center. This must be true. Now, suppose we had a circle but mislaid its center – how would we find it? In Euclidean geometry, it's easy: we would construct the perpendicular bisectors of two chords, which would meet at the center. The perpendicular bisector of a chord contains a diameter. So, what we really need are hyperbolic diameters of the circle. It is well past time for a decent picture.

2.2 Finding the hyperbolic center

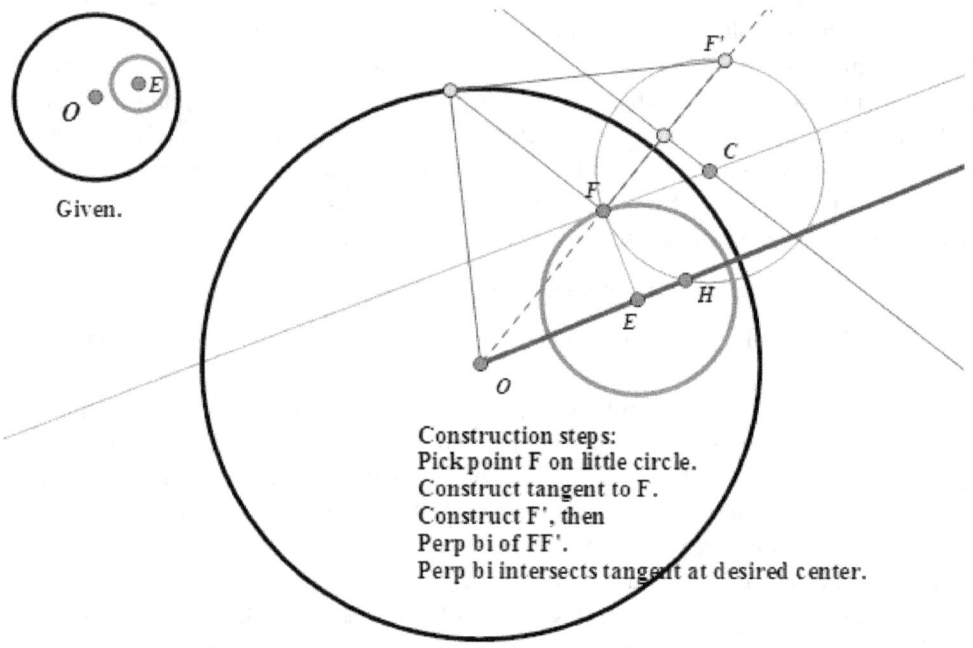

Given.

Construction steps:
Pick point F on little circle.
Construct tangent to F.
Construct F', then
Perp bi of FF'.
Perp bi intersects tangent at desired center.

The small picture in the corner shows the situation: we have a hyperbolic circle and its Euclidean center E. The larger figure reveals the location of H, the hyperbolic center. We have to understand why this construction works. We need hyperbolic diameters, right? The diameter through $\overline{OE}$ is one. Now all we need is one more hyperbolic line orthogonal to the little circle.

We have to solve a case of double orthogonality: we need a Euclidean circle orthogonal to the boundary and to the little circle. Since any hyperbolic line orthogonal to the little circle will do, we pick the point F on the little circle and construct the tangent at F. The construction marks are Euclidean segments $\overline{FE}$ and $\overline{FC}$. The center of the circle we seek has to be on the tangent to the little circle in order to be orthogonal to the little circle.

Then we construct the inverse of F, designated $F\prime$. Since we want a hyperbolic line through F, we construct its inverse $F\prime$, (the thin lines through F and $F\prime$ in the figure are the construction marks). The Euclidean perpendicular bisector of $\overline{FF\prime}$ intersects the tangent line at the desired center. The red circle C contains the hyperbolic line through F, orthogonal to the little circle. Therefore it contains a hyperbolic diameter of the little circle. The two hyperbolic diameters meet at H, the hyperbolic center of the circle.

The point F was chosen at random, so that point may be moved around the little circle. So long as F is chosen conveniently, the hyperbolic center would always turn out to be in the same place. For those who try this drawing on paper, choosing F so that the construction stays in the space provided requires some foresight. Making the figure in *SketchPad* allows dragging the little circle toward the boundary: worth the effort just to see what happens.

Before we finish considering the axioms, this is a good time to look at an application of the hyperbolic center. Suppose we have a hyperbolic segment and we want its hyperbolic midpoint?

The figure below contains all the necessary thoughts; so an intrigued reader should turn away for a minute and think about it. The hyperbolic midpoint cannot be something as simple as the Euclidean midpoint of the Euclidean arc because the proximity of the endpoints to the boundary will decide the length of the segment, and the location of its hyperbolic midpoint.

The construction is simple. We use the straightedge first to draw the radii of the hyperbolic line to the endpoints of the segment. Then we construct their tangents, which intersect at point E. Finally, the ray $\overrightarrow{OE}$ intersects the given segment at its midpoint M. (The construction does not need the little circle inside the tangents – that circle is used for the proof, below.)

2.3   Hyperbolic midpoint

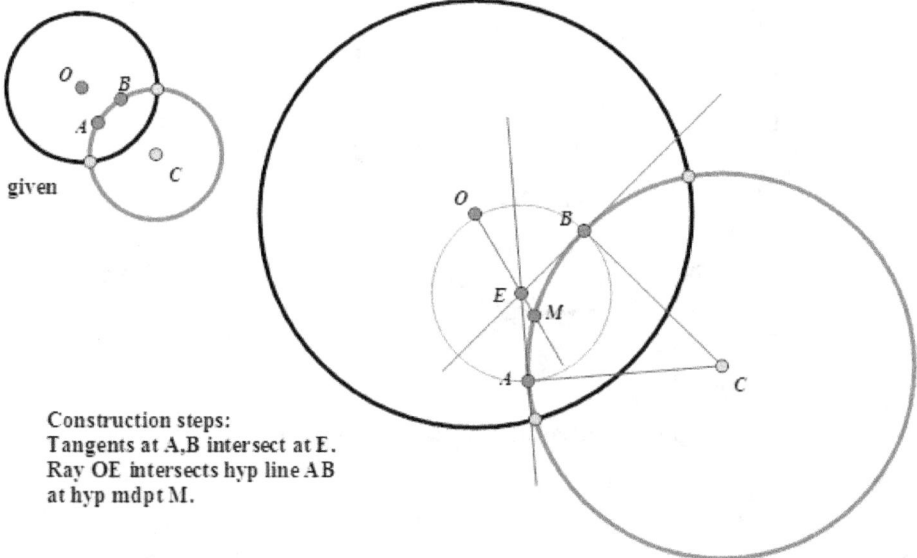

given

Construction steps:
Tangents at A,B intersect at E.
Ray OE intersects hyp line AB
at hyp mdpt M.

Proof: The proof that the construction works uses the circle with Euclidean center E through the endpoints of the segment. We note that the original hyperbolic segment $\overline{AB}$ lies on a hyperbolic line, orthogonal to the thin circle E. The segment thus contains a diameter, as does the ray $\overrightarrow{OE}$. Therefore the ray and the segment intersect at the hyperbolic center. This center must be hyperbolic equidistant to any point on the circle, making the point M the hyperbolic midpoint of the diameter $\overline{AB}$. ∎

To finish verifying the axioms are true in the model, we only have to consider right angles and the situation with the parallel lines. Right angles are all congruent in hyperbolic geometry because we measure the angles of curved objects with tangents. Orthogonal circles will have perpendicular tangents, so all our right angles, even curvy ones, are the same size.

Now, we need to establish the negation of the usual parallel postulate: through a point not on a hyperbolic line, there exist more than one line parallel to the given line. We can think of a hyperbolic line easily now. And, for some point not on it, we can construct hyperbolic lines which miss the given line. The situation resembles the figure below. We observe lines which completely miss the given line by some space we can see. (Construction marks have been included.)

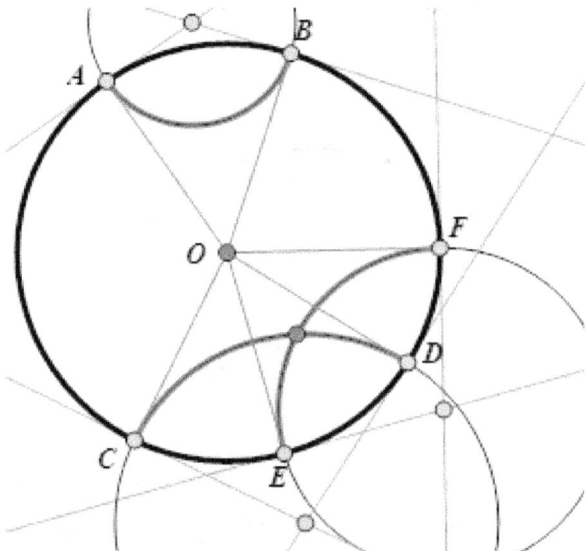

Negation of parallel postulate verified.

But a student thinking about the subtleties of the disk might consider another case, one where two hyperbolic lines meet on the boundary. Since the boundary points are not in our space, the hyperbolic lines do not meet. But they are as close as they can get without meeting.

We name hyperbolic lines which meet on the boundary *limiting parallel.* They will end up being good friends of ours, with easy constructions and useful applications. They will also help us understand the full story for properties like the sum of angles of a hyperbolic triangle because the limiting parallels represent the limiting case, the unattainable yet bounding version of what can be done in the hyperbolic disk.

We name hyperbolic lines which intersect neither in nor on the boundary as *ultra parallel.* These are the parallels with some space between them at both ends. In the example below, hyperbolic lines $\overleftrightarrow{CD}$ and $\overleftrightarrow{AJ}$ are ultra parallel with $\overleftrightarrow{OA}$ and $\overleftrightarrow{AJ}$ limiting parallel.

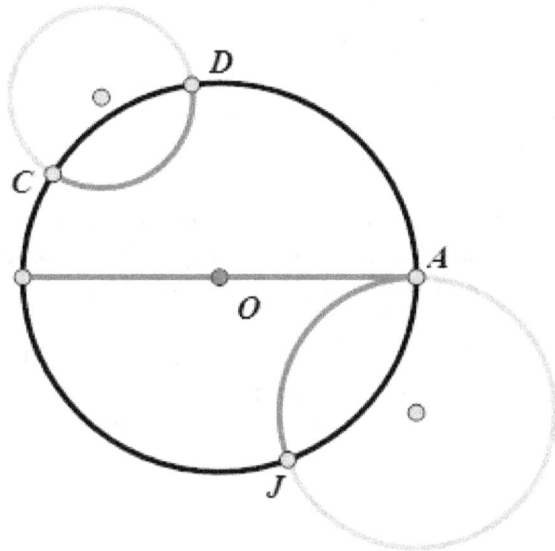

Ultra parallel and limiting parallel lines.

So now we see that it is possible to find an interpretation of the first four Euclidean axioms with isometries and a negation of the parallel postulate all true at the same time. This model, along with Euclidean geometry, implies that the parallel postulate cannot be proved from the rest of our assumptions.

This is easily seen. Suppose we could prove the parallel postulate from the first four axioms and the isometries. Then, within the hyperbolic interpretation we could prove the parallel postulate. But we already have the negation of the parallel postulate true. This would give us a contradiction: a statement and its negation true at the same time.

Working in an axiomatic system which allows contradictions would be worthless because we could prove anything. In such a system, if $2 = 3$, then $1000 = 57 = 3.14$. Literally any statement would be true. There would be no point in working with such a wishy-washy system.

Putting the hyperbolic model inside the Euclidean model is particularly clever here because it means that Euclidean and hyperbolic geometries have the same consistency – any contradiction in one implies a contradiction in the other. So far, through all the years, neither geometry appears to contain a contradiction. All the math people work with both geometries as if they are consistent. (It's not like we have any choice!)

We conclude that the parallel postulate is independent of the other four axioms and the isometries. This knowledge was unknown to mathematicians from the time of Euclid through the mid 1800's. Many people consider this discovery of the independence one of the biggest intellectual achievements in math.

The above discussion is included in college geometry and math history courses all over the world because it settles an ancient, difficult problem in a surprising and intellectually rich way. The disk model's intellectual potency settled some hard questions which remain some of the most important questions of the book.

- Why is the parallel postulate independent of the other axioms and isometries?
- What is the logical standing of hyperbolic and Euclidean geometries?
- What theorems would be true in both Euclidean and hyperbolic geometry?

We will address these questions in homework and in the pages to come. Understanding their answers connects us with some powerful ideas and puts us ahead of some very bright men and women who lived just a few centuries ago.

2.4   Angle sum of a hyperbolic triangle

The first people to explore hyperbolic geometry, before it was even known to be a geometry, reasoned to many results which, at the time, seemed weird. But weird is not necessarily wrong. One of the key, and weirdest, properties of hyperbolic triangles is the sum of their interior angles. We will now construct an example to stand for all cases which shows that the sum of the angles of a hyperbolic triangle is less than 180 degrees. This surprise was found before the disk model existed and some mathematicians took it as an absurdity, a contradiction that implied the negated parallel postulate led to nonsense. But when we see how the sides of a hyperbolic triangle are bent, we can see the sum is indeed less than 180 degrees. And we can see much more.

So, let us begin with a hyperbolic triangle. This means each of its sides must lie on a hyperbolic line. Some interesting shapes are possible, as we shall see later. Now, we can use translation to slide this triangle so that one of its vertices lies on O. If we were to see a cartoon of this happening, our Euclidean eyes would be surprised to see that the two sides meeting at O would straighten out into Euclidean segments. The triangle would have to be congruent to the original because all we did was translate the triangle. But remember, in the disk, distances change relative to the proximity to the boundary. So the triangle might look different after translation, but it is in fact exactly the same triangle in hyperbolic geometry. But with two Euclidean straight sides, the angle at O can be measured without tangents. The other two angles have a side lying on a hyperbolic line which has to be part of a Euclidean circle. We must measure those angles with their tangents. The figure below gives the situation. The reader should see that the angle sum is less than 180 degrees. In general, the sum must be less than 180 degrees because the tangents at A and B must meet between the Euclidean segment $\overline{AB}$ and O. The only way for the angle sum to be 180 degrees would be for the third side to be straight. That cannot happen unless the entire triangle approaches O and degenerates to a single point.

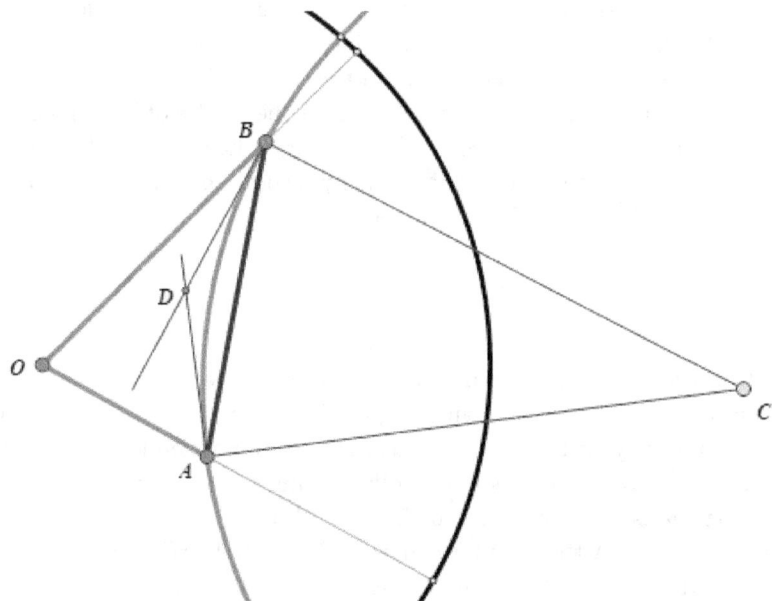

Angle sum of a hyperbolic triangle.

In the figure above, the hyperbolic triangle has sides $\overline{OA}$ and $\overline{OB}$, which are Euclidean straight, and $\overline{AB}$, which is also a Euclidean arc. So the triangle looks like a spear tip or shark fin. The Euclidean angles OAD and OBD illustrate how to measure the hyperbolic angles using tangents and we don't even need specific sizes. The big idea is, the tangents always meet inside the hyperbolic triangle while the triangle's third Euclidean side is outside the hyperbolic triangle. Therefore the angle sum of any hyperbolic triangle has to be less than 180 degrees.

The fact that the angles of a hyperbolic triangle always add up to an amount less than 180 degrees might seem like big news, but there's much more we can reason from this key result. For instance, what if we consider the sum as the sides approach limiting parallel? Since limiting parallel lines share the same tangent line, the angle between limiting parallel lines is zero. Yes, we cannot have a hyperbolic angle of zero because the vertex would not be in the space. But we arrive at the mundane conclusion that angles have a lower limit size of zero with the stunning observation that angles with measure zero have a region of space inside. A Euclidean angle snaps shut as the measure approaches zero. This still happens in hyperbolic space when the vertex of the angle is inside the boundary. But hyperbolic space is indeed strange: an angle whose measure is approaching zero may or may not snap shut, depending on whether the vertex of the angle approaches the boundary or not!

The angles formed by limiting parallels are called asymptotic angles, a suitable name due to their status as a limiting case, just like asymptotes in graphs. The reader will be asked to construct a triangle with three asymptotic angles, called a triply asymptotic triangle. It represents a limiting case at each vertex, with each angle approaching zero. This finishes the story of the sum of the angles of a triangle. The upper limit in hyperbolic geometry for the sum of the angles of a triangle is 180 degrees, and the lower limit is zero. Neither limiting case has an actual hyperbolic triangle with those angles sums. Remember, in hyperbolic geometry, we have no fixed sum of interior angles, as in Euclidean geometry.

2.5   AAA

The upper limit on the angle sum of a hyperbolic triangle gives us another surprise: if two hyperbolic triangles have the same shape, then they are congruent. Chalk up another weird property of hyperbolic space: no similar triangles. Considering how much mileage we have gotten out of similar triangles in Euclidean geometry, hyperbolic geometry must be very different from Euclidean.

We will call this our AAA theorem for congruent hyperbolic triangles. The proof is simple enough – another proof by contradiction.

If two hyperbolic triangles have three pairs of congruent angles, the triangles are congruent.

Proof: Suppose we had two hyperbolic triangles which have the same shape but different sizes. So they have three congruent pairs of angles and one of the triangles is bigger than the other. We can use isometries to put the smaller triangle inside the larger one one, lining up the vertices of one of the angles so that the sides also align. We may as well put the corresponding angles in corresponding positions.

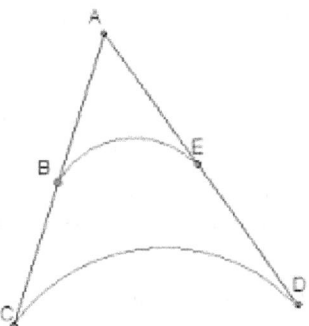

The above figure has two sides looking straight and they certainly don't have to be. The shape of the sides makes no difference because the source of the contradiction we seek lies before us. Remember, we measure hyperbolic angles using tangents. So a pair of adjacent hyperbolic angles with their exterior sides on a hyperbolic line form a pair of supplementary angles. That means, in our figure, the adjacent hyperbolic angles ABE and EBC are supplementary, as well as angles AEB and BED Then the hyperbolic quadrilateral $CBED$ has angle sum 360 degrees. That sounds Euclidean, which is impossible.

Since there exists a hyperbolic line through C and E, we can cut the quadrilateral into two hyperbolic triangles, each of which must have angle sum less than 180 degrees. Thus, we cannot have the sum of all four angles equal to 360 degrees. ■

2.6 Mutual perpendiculars

So, if two hyperbolic triangles have their three angles pair-wise congruent, the two triangles are congruent. They may well look different in the Euclidean sense. But in hyperbolic space, they are congruent. This leads some geometers to say things like, "there is only one hyperbolic regular pentagon with all right angles." This is because two regular hyperbolic pentagons might look different, but their angles are all the same, so they must be hyperbolic congruent. We will find an easy construction for such objects and observe how different they may look in Euclidean eyes. This construction depends on another surprise: two ultra parallel hyperbolic lines have exactly one hyperbolic line perpendicular to both lines. In Euclidean space, we can have many mutual perpendiculars to a pair of parallel lines. They look like ladders. We see, from the proof we just did, that hyperbolic space has no ladders, that ultra parallel lines have a unique mutually perpendicular line.

Now, if there is a unique mutual perpendicular, we had better be able to construct it. The situation is this: we have two ultra parallel hyperbolic lines. One could be a diameter (this case is left for homework.) We seek the hyperbolic line which is perpendicular to each. The puzzle sounds tricky: we need to solve a case involving three orthogonal circles, the two lines and circle O, looking for a fourth circle orthogonal to these three.

We can construct a hyperbolic line perpendicular to another hyperbolic line at a specific point without much trouble. This is left as a homework problem. Then we can construct a chain of perpendicular hyperbolic lines and look for a pattern, a connection to use when we have ultra parallels. It would be as if we are missing a link in the chain. So, finding a new construction might happen when we start with a situation with the given information and solution already present. Then we pretend the piece we are trying to find is missing and we look for ways to find the missing piece from the given information. This strategy is particularly good for our problem because we can build the pieces as described in this paragraph and see that the secant line through the endpoints of the hyperbolic lines gives us an amazing shortcut.

If two circles intersect in two points, the line through those points is called the radical axis. In the figure below, the two radical axes are thin, intersecting at point M. This point M is the center of the circle containing the mutual perpendicular. That was fast. Two quick, Euclidean lines built from the given and we have our desired center already! In fact, we will prove that any hyperbolic line perpendicular to a given hyperbolic line must have its center on the radical axis formed from the given line and circle O. The reader should keep this fact in mind because we will use perpendicular hyperbolic lines quite often in this section. For instance, we will be building polygons with all right angles: radical axes will be crucial. But before we prove our construction works, let's take a look at the construction itself.

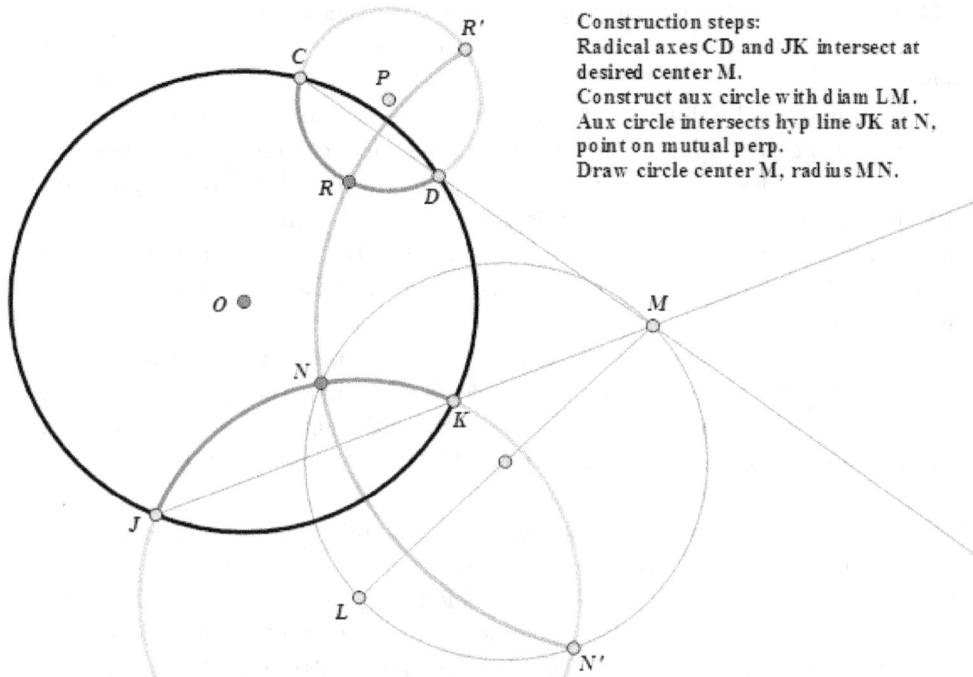

Construction steps:
Radical axes CD and JK intersect at desired center M.
Construct aux circle with diam LM.
Aux circle intersects hyp line JK at N, point on mutual perp.
Draw circle center M, radius MN.

The unique mutual perpendicular.

The given circles are centered P and L. The radical axes meet at M, which we find with just a straightedge. Now we choose one of the two centers of the given hyperbolic lines and draw the segment from M to that center, the segment $\overline{LM}$ below. We next construct the thin auxiliary circle with diameter $\overline{LM}$. This circle intersects the given hyperbolic line at N. The point N lies on the mutual perpendicular, so the circle with center M through N is the unique mutual perpendicular.

The figure has many more points labeled because we will use the same figure for the proof. So much further work relies on this construction that we must regard it as required knowledge.

Proof: We will use the construction of the mutual perpendiculars within the proof. We will use the same figure as the construction instructions. We are given the thick circles, the ones orthogonal to the circle with center O. We first seek to prove that the point M, where the radical axes meet, is the center of the unique circle orthogonal to all three given circles.

So long as the given hyperbolic lines do not have Euclidean parallel radical axes with circle O, the radical axes will intersect at M, as pictured. (The case with parallel radical axes is in the homework.) We then choose one of the centers of the given hyperbolic lines, L in the figure, and construct the Euclidean segment $\overline{LM}$. Then we use this segment as a diameter and construct its circle. The undrawn Euclidean radii $\overline{LN}$ and $\overline{MN}$ are perpendicular, for a reason we leave as a homework problem.

We then construct the thin Euclidean circle with center M and radius $\overline{MN}$. The last sentence of the previous paragraph gives us the circle centered at L orthogonal to the circle centered at M. The radical axis through J and K now means that J is the inversion of K across the circle centered at M. Since J and K are on the boundary, the circle passes through a point and its inverse, making circle O orthogonal to the circle centered at M. This means that the circle centered at M qualifies as a hyperbolic line.

The radical axis $\overline{CD}$ now means that C is the inverse of point D across the circle centered at M. (We are thinking of circle O as the hyperbolic line in the disk with center M, but only for a minute.) The circle with center P passes through a point and its inverse, so that circle must be orthogonal to the circle centered at M.
■

Although the proof is subtle, the construction is simple, with an unusually large reliance on the straightedge. With this theorem and the homework problems, we can create polygons with all right angles, as long as they have more than four sides. A new result regarding regular constructible hyperbolic polygons is also on the horizon: we will know which all right-angled regular hyperbolic polygons are constructible. Because of the flexibility of interior angle sums for hyperbolic polygons, there exist regular hyperbolic pentagons with all angles 100 degrees and all angles 0 degrees. . . any size which keeps the angle sum less than the Euclidean angle sum.

Another fact to keep in mind is, the all right-angled polygon does not necessarily have all sides congruent unless we make them so. Here is an example of a hyperbolic hexagon with all right angles which, by appearances, does not look regular.

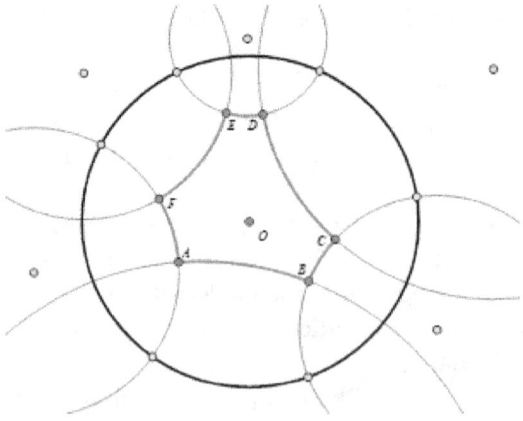

2.7   Miquel's Six Circles and more orthogonal circles

Miquel's Six Circle Theorem appears in some geometry texts and recreational mathematics books. So a student who liked the Miquel point in the first section might find this theorem after searching for further work from Auguste Miquel. The picture which accompanies the statement of the theorem suggests connections with the orthogonal circles we just saw because, maybe, the outer circles could all be orthogonal to circle O. We can see that they are not orthogonal in this figure because center B is inside circle O. Orthogonal circles have their centers outside each other.

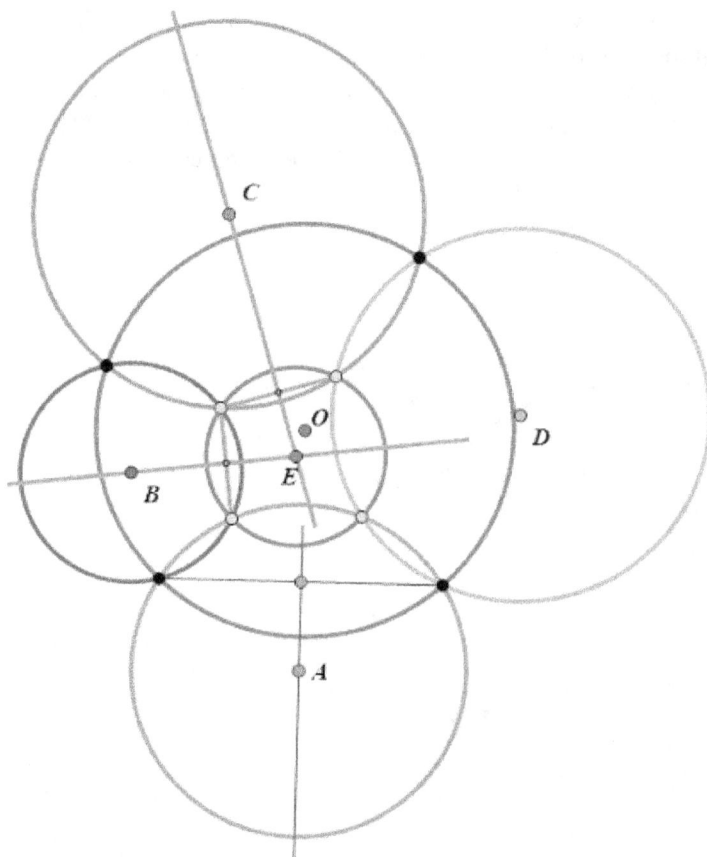

Miquel's six circles.

Given points A, B, C, D on a circle, construct a circle through each consecutive pair of points so that neighboring circles intersect twice. Then the four intersections of neighboring circles lie on one circle.

It should be noted that this sixth circle could be inside the first circle or outside it, depending on the placement of the points A, B, C, D and the circles chosen.

Let's get creative. The six circles can be related in some interesting patterns. Now, suppose the reader seeks an active mathematical experience. He or she might look for a connection with something familiar, like orthogonal circles.

Let's play with this idea: if we can't get four neighboring mutually orthogonal circles, what if we had three? For those trying to make their own drawings, here is a method for constructing two orthogonal circles cheaply. Start with two points on one circle and construct the tangents at those two points. As long as those two points are not diametrically opposed, their tangents will intersect. If we use that point as the center and either of the two given points as a point on the circle, we can draw a second circle whose radii contain these two tangents. Here is a figure showing some steps in the thought process toward our new idea. We are playing with the four orthogonal circles, three of them in a chain along the perimeter of the fourth. The sets of points which were on one circle in the Miquel Six Circle Theorem appear to be on circles as well in our figure. More specifically, the innermost four intersections lie on a circle and the outermost four intersections appear that they could lie on a circle as well.

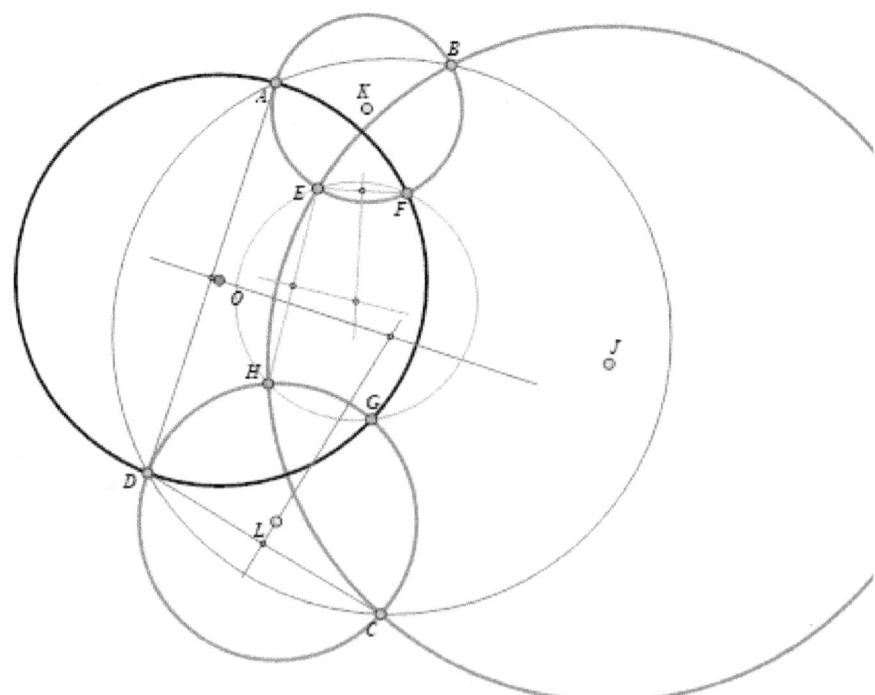

Our orthogonal circles.

2.8 Our orthogonal circles conjecture

So here we are, with an idea that looks interesting. We have constructed a figure which contains the given information; the circles have been built to be orthogonal. (We will learn how to other ways to construct such figures in Section 2.) We then constructed the circle through points E, H, and F and this circle appears to catch point G. Of course, appearing to pass through all four points is not the same as proving these four points lie on one circle. But before we invest time thinking about the general case, we should study with the drawing. Ideas: Four points on a circle: cyclic quadrilateral. Orthogonal circles: perpendicular radii.

This is compelling evidence that we have a truth in our reach. What must we do to find proof of the general case? Perhaps being orthogonal is not even required! Maybe four circles arranged as we have them will have their four inner intersections lying on a circle. It is possible that we are looking at a special case of a more general theorem. The more general theorem would be more powerful, if it were true. So here's a quick look at a more general case, which fails to work. The points B, C and D were used to construct a circle and the circle misses point A. So, orthogonality to circle O and each other seems to be required.

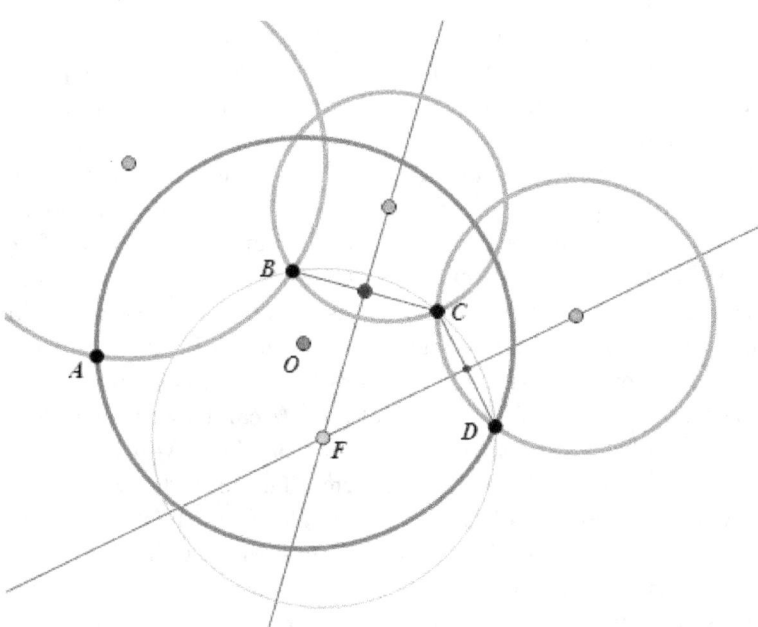

This chain of circles fails our weakened conjecture.

So we are back to our original conjecture, with the further knowledge that being orthogonal appears to be important. This is most useful knowledge because it means, out of all the segments we could draw, we must include the radii ending at points of intersections. We could spend a frustrating hour looking at other choices of segments which don't work out. But using the essential properties of the given information has to be our first attempt because it is most logical to build on required structure rather than coincidences. Let's state our conjecture.

Given four mutually orthogonal circles so that three are orthogonal to one and these three are orthogonal neighbors, their four outer intersection points lie on one circle as well as their four inner intersections.

Writing the conjecture in a clear manner is a crucial step in creating mathematics because its formation is the official start of the proof. We always want the best start we can get. Next, we want a good figure to use. As mentioned on the previous page, we will make our first attempt at the proof using the perpendicular radii which are hidden in orthogonal circles.

Unlike Miquel's six circles, we do not have four circles in a chain around one circle, no matter how we look at it. The reader is urged to study the ideas and the "Our orthogonal circles" figure again. A strategy should be forming. Take some time to write down what it might be. The reader should make a pencil sketch and mark the right angles. Are there any other angles known to be equal? Mark them down. Keep the goal in mind: we want one pair of opposite angles in that quadrilateral to be supplementary.

We should stop here a while and enjoy the experience. We are pretty sure the conjecture is true. We have a good drawing with lots of information. There must be a way to finish the process, to show that our given information forces the quadrilateral $EFGH$ to be cyclic. Seeking the answers is much better than simply following along. And finding the proof unaided is an intellectual accomplishment.

Proof: There are four isosceles triangles in the figure below and each has a circle's center as a vertex. The triangle interiors have been shaded. The strategy for our proof is this: each angle of quadrilateral $EFJH$ can be written as 90 degrees plus or minus a pair of angles from the isosceles triangles. We will add the four angles of the quadrilateral in this expanded form. The four 90's will cancel their sum of 360, leaving a signed sum of all the base angles from the isosceles triangles. Since the base angles are equal, we divide by two and see that there is a combination of base angles which is zero. When we add a pair of opposite angles in quadrilateral $EFJH$, we get 90 plus 90 plus this same combination of four base angles. So the sum is 180, verifying that the opposite angles are supplementary. The same strategy works for the outer four points of intersection.

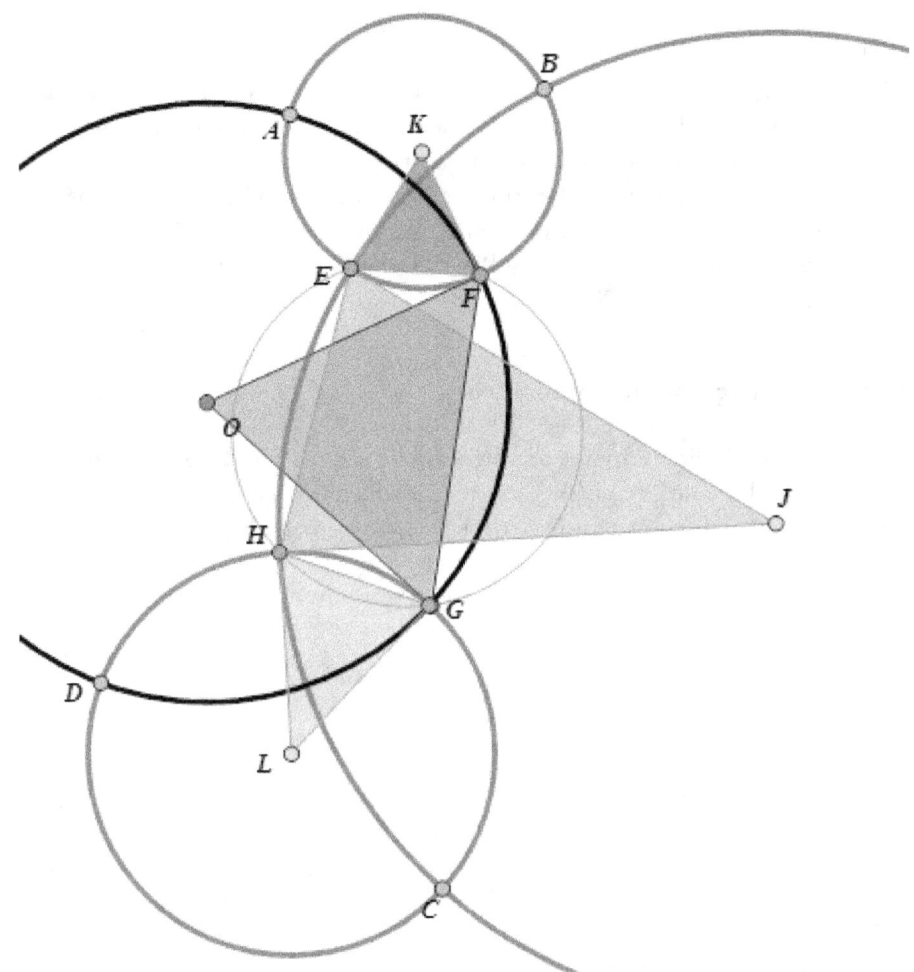

Our orthogonal circles proof.

First, the sum of the angles of the quadrilateral $EFGH$ can be written as $\angle EHG + \angle HGF + \angle GFE + \angle FEH$. We now use those perpendicular radii.

$$90 - \angle LHG + \angle EHJ + 90 - \angle LGH + \angle OGF + 90 - \angle KFE + \angle OFG + 90 - \\ \angle KEF + \angle JEH = 360$$

Each triple of letters occurs twice, in different order. Each of these pairs are base angles of an isosceles triangle, so we can substitute ($\angle OGF = \angle OFG$) and rewrite. Division and sums of adjacent angles give us:

$$2(90 - \angle KFE + \angle OFG + 90 - \angle LHG + \angle EHJ) = 360.$$
$$\angle EFO + \angle OFG + \angle GHJ + \angle EHJ = 180.$$

$$\angle EFG + \angle EHG = 180.$$

Now, the signs on the sum depended on the centers being outside the quadrilateral. When we examine the outer four intersections, some of the centers could be inside the quadrilateral, or even on some of the sides. We don't have to worry too much about all the cases because the substitutions we use in the sum of all four angles would have to be the same substitutions we would use when adding a pair of opposite angles. So the adjustments in the 360 sum will add to zero. Considering the sum of a pair of opposite angles would have to use the same sum of adjustments, plus two 90 degree angles.   ■

2.9 Hyperbolic perpendicular bisector

We can find the hyperbolic midpoint of a hyperbolic segment and we can construct perpendicular hyperbolic lines quickly as well. This means we can construct the hyperbolic perpendicular bisector of a hyperbolic segment. Let's look for the quickest way, using our best moves.

It might look like a busy construction. But there are a lot of Euclidean lines, which are always easy to make.

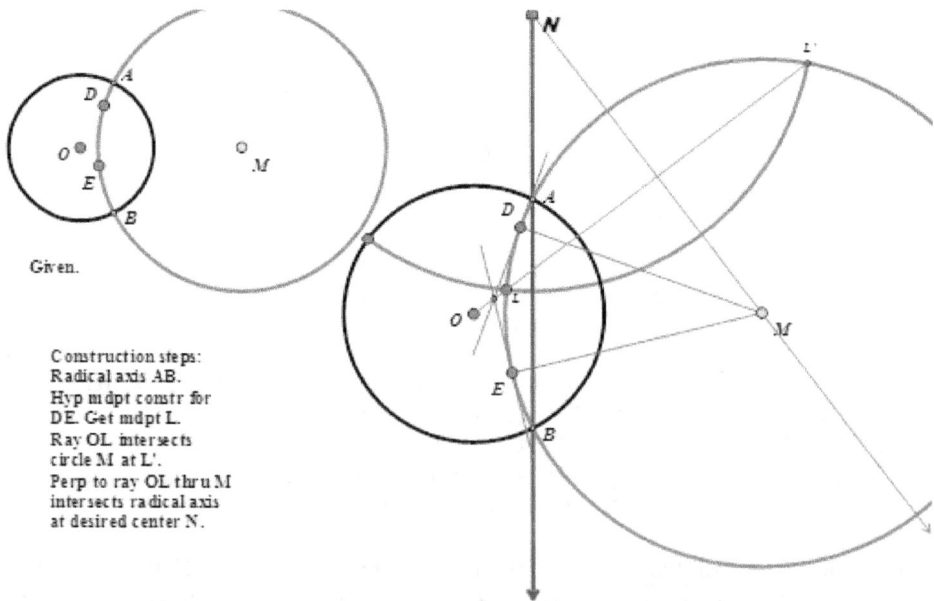

Given.

Construction steps:
Radical axis AB.
Hyp midpt constr for
DE. Get mdpt L.
Ray OL intersects
circle M at L'.
Perp to ray OL thru M
intersects radical axis
at desired center N.

Proof: The Euclidean segments tangent to circle M at D and E start the hyperbolic midpoint construction, giving us L, the hyperbolic midpoint of hyperbolic segment $\overline{DE}$. The Euclidean line $\overleftrightarrow{PM}$ is the perpendicular bisector of the Euclidean segment $\overline{LL\prime}$. So, the circle P goes through a point and its inverse, making it a hyperbolic line. Its Euclidean center is on the radical axis AB, so circle P must be orthogonal to circle M Therefore, the hyperbolic line $\overline{LL\prime}$ must be perpendicular to the given hyperbolic segment, through its midpoint L. ■

2.10 Asymptotic triangles

We have mentioned triply asymptotic triangles a few times. It's about time we learned to build this limiting case of a hyperbolic triangle whose angles are all zero. We are going to write as if the triangle is an entirely hyperbolic entity, even though its vertices are not in the space. After all, we could build a hyperbolic triangle as close to this limit as we want, so we may as well act as if the triply asymptotic triangle is in the family.

Hyperbolic lines, not segments, form the three sides of the triply asymptotic triangle ABC. The construction is left as a homework problem with the hint here that the Euclidean centers for the sides of an asymptotic angle lie on one tangent. All the triply asymptotic triangles are congruent. We shouldn't say this is so by AAA, since the asymptotic triangles are not entirely in the space. However, we can still note that their angles are all zero and the sides are all hyperbolic lines: so each triply asymptotic triangle has to be the same size and shape as the others.

Here's an example of a triply asymptotic triangle ABC. Like the hyperbolic distance formula, we will not prove by construction how to calculate the area of a hyperbolic triangle. It turns out the area of any hyperbolic triangle in circle O is a constant for the radius of circle O times π minus the sum of the angles of the triangle. We will use 1 for the constant, so the area of a hyperbolic triangle can be written

$$A = \pi - \alpha - \beta - \gamma$$

This formula implies that the triply asymptotic triangle has the upper limit for triangle area: π. Hyperbolic triangles can be constructed in many ways; but their area has an upper bound.

Here's a brief digression using limiting parallels in Euclidean geometry. Suppose we have two circles which are tangent and outside each other. How would we construct their mutual tangent? This would make a good problem to mull over with compass and straightedge because of the simple given. In the figure below, the given circles are circles O and A.

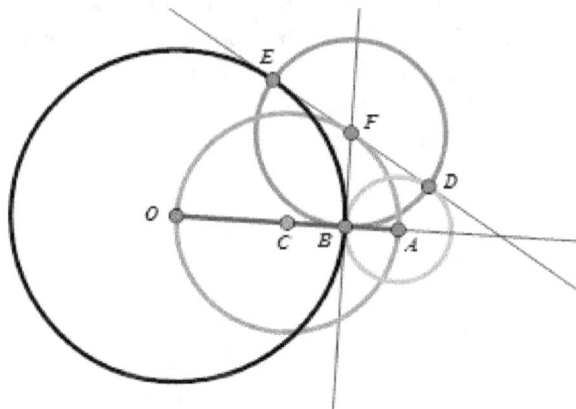

Mutual tangent.

The proof is not too bad, so we will leave it for the next homework assignment. In that problem, we will see that circle F is orthogonal to circles O and A. We also have $\overline{DE}$ as diameter of circle F, so asymptotic triangle EDB is triply asymptotic in circle F.

That last problem showed a surprising relationship between a Euclidean construction and a triply asymptotic triangle. Our next excursion shows the connection between Euclidean and hyperbolic regular polygons. In Section 1, we briefly mentioned that some regular Euclidean polygons cannot be constructed with a compass and straightedge. But we recently found out that we can construct things in hyperbolic geometry which would have been impossible in Euclidean geometry. For example, we could construct a hyperbolic hexagon with all angles 90 degrees and we could construct many lines through a given point parallel to a given line. We have seen hyperbolic circles whose hyperbolic centers are not their Euclidean centers. So, something impossible in Euclidean geometry might be possible in the hyperbolic disk.

The first thing we will do is to start with a regular Euclidean polygon (with 5 or more sides) and show how to get a regular hyperbolic polygon with the same number of sides and all right angles. We will do an example starting with a regular, Euclidean hexagon. First, we will prove that our hyperbolic construction works.

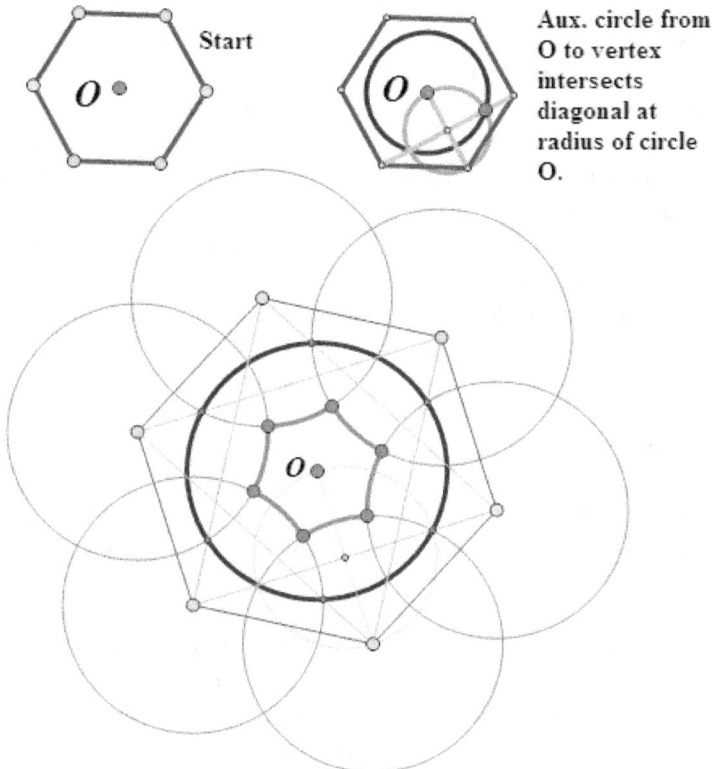

Circle O intersects diagonals which skip one vertex at points on boundary for hyperbolic lines. Use vertices of hexagon as centers.

Construction for regular hexagon with all right angles.

Proof: We will describe the construction and explain why it works as we go. First, we draw in the diagonals which skip one vertex. All we have to do next is illustrated in the middle drawing: construct the circle whose diameter is the segment from the center to a vertex. The diagonal which skips the vertex we used intersects this circle twice. These two points are on circle O. Yes, we did not start with the boundary in the picture; but we place it wisely and we get all the hyperbolic lines we want. (This is a very cool trick to keep in mind.) Angle OFC is a right angle because it is inscribed in a semicircle. This guarantees the red circle with center at C is a hyperbolic line in circle O. Using the diagonals we drew as radical axes, we get hyperbolic lines which are perpendicular to each other. The last figure has the desired hyperbolic hexagon marked with grey vertices. This construction never relied on the number of sides of the given polygon, so it will work for any given polygon. ■

On the next page, we will find out why we cannot construct an all right angled hyperbolic heptagon. Aquinas student Jillian Duffey used these ideas in her 2009 undergraduate research and published her results in the *Rose-Hulman Undergraduate Math Journal*. The work on the previous pages guarantees that we can construct any all-right angled regular hyperbolic polygon, provided we have the Euclidean version given. (See Jillian's article for an alternate construction for such polygons.) It is known that some regular Euclidean polygons are not constructible. For instance, the seven-sided regular polygon, or regular heptagon, is not constructible. So, our next quest concerns the hyperbolic versions of impossible Euclidean polygons. We will now prove that such a construction does not exist.

2.11   Jillian's Theorem

The number of sides of constructible, all-right angled regular hyperbolic polygons is exactly the same as constructible regular Euclidean polygons, for $n > 4$.

Proof:

Suppose we could construct an all-right angled regular hyperbolic heptagon. (Here heptagon, seven sides, stands for any Euclidean, non-constructible number of sides.) If the heptagon is built with its center at O, we could join the adjacent vertices with Euclidean segments and we would then have the Euclidean regular heptagon, by construction, which is impossible. Therefore, the only way we could have a construction for the hyperbolic heptagon would be to build it off-center.

Even if we, for some bizarre reason, can construct this heptagon off-center but the construction cannot be done on center, we have constructed the heptagon. Some of the sides will have different Euclidean arc lengths from each other, if the reader is trying to imagine this heptagon. There exists only one regular heptagon with all right angles: this hyperbolic side length is standard in circle O. (Remember, in hyperbolic space, similar polygons are congruent.) All we have to do is translate the heptagon so that its hyperbolic center lands at O. Good news: translation is an isometry which is constructible. (We will have a few hyperbolic isometry constructions in the homework.)

Well, then, here we are with the vertices of the hyperbolic heptagon evenly spaced around O. All we have to do is join consecutive vertices with our ruler and we have accomplished that which has been proved impossible: we constructed a Euclidean regular heptagon. That cannot happen. Therefore, no construction exists for the hyperbolic, regular, all right-angled heptagon. We used the heptagon to stand for any of the non-constructible Euclidean regular polygons. We certainly never referenced the number of sides in any important way, so our theorem holds for $n > 5$. ■

2.12   Noah's construction

We can make angles of all sorts of sizes in hyperbolic geometry. In fact, the constructible angles are the same as Euclidean, since we measure angles with tangents. In degrees, sizes like 30, 45 and 60 are easy to make. Adding, subtracting and bisecting give us other sizes. There are also the angles we can get from right triangles made with sides of constructible lengths. For example, we can construct an angle of size $\arctan(\frac{3}{4})$ by constructing a 3,4,5 triangle. Not only does $\arctan(\frac{3}{4})$ fail to have a nice degree size; but in radians it turns out to be an irrational multiple of π. (It takes complex numbers to prove this. The easiest proof is in <u>The Book of Numbers</u>.) Our list of constructible angles in radians has some gaps, like $\frac{\pi}{7}$. (The reader can see that constructing an angle of $\frac{\pi}{7}$ would give us a regular Euclidean heptagon pretty easily.)

These angle ideas are tied to a famous math problem. In the Euclidean section, we read that constructing a square and a circle of the same area is impossible by construction. When Bolyai wrote up his hyperbolic geometry studies, he closed with a result meant to catch the eye of mathematicians: squaring the circle can done for an infinite number of cases in hyperbolic geometry. Unfortunately for him, almost nobody noticed.

Aquinas student Noah Davis found a construction for squaring the circle in the Poincaré disk, which he used to prove that Bolyai's strategy for this construction was the only possible strategy: we must build the square and circle separately. (Another mathematician, W. Jagy, had the logic result in 1995, but he did not have the actual construction. Jagy proved that construction exists without finding it.) Noah's work starts with a given or constructible angle and builds a regular hyperbolic quadrilateral with four angles of that size. It's interesting that he builds circle O after some Euclidean work.

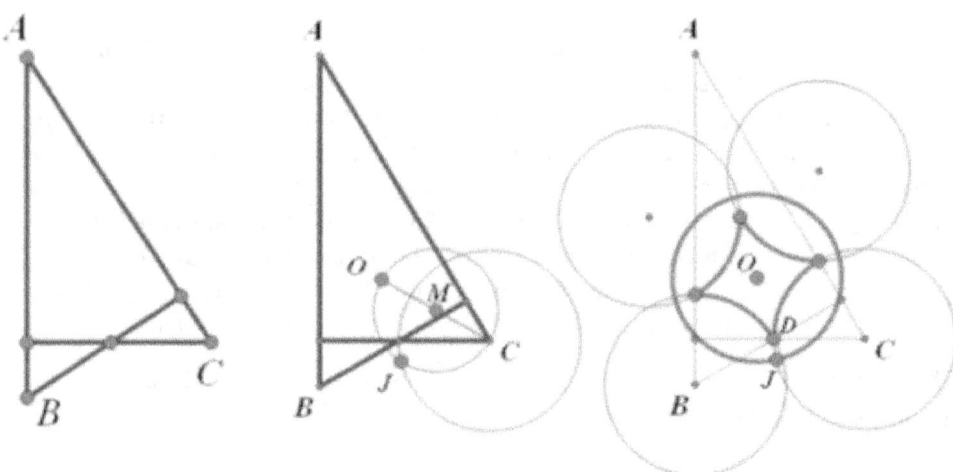

Construction of a regular hyperbolic quadrilateral with vertex angle A.

Angle A is the desired angle, in this case, $\dfrac{\pi}{6}$. Two congruent right triangles are built as pictured. The point O is the incenter of either right triangle (same location for both.)The auxiliary circle M and circle C as pictured determine the size of circle O. Circle C can be reflected three times to finish the job. (We leave the proof that this construction works to Noah's article.)

The careful reader might feel ripped-off because we promised to square the circle and all we made was a regular hyperbolic quadrilateral with all angles $\dfrac{\pi}{6}$. Since we can't have four 90 degree angles in one hyperbolic quadrilateral, a Euclidean square isn't possible. And since the Euclidean square is a regular quadrilateral, we will call our regular quadrilateral a square, even though it's hyperbolic, not Euclidean. Since lots of other angle sizes are possible, perhaps Bolyai noticed something wild about hyperbolic geometry: it has squares of all angles sizes between 0 and 90 degrees, many of them constructible.

The hyperbolic area of this square is $2\pi - 4(\dfrac{\pi}{6}) = \dfrac{4\pi}{3}$. In order to find a circle of the appropriate size, we need to find the radius. We'll use a formula known to Bolyai, without proof.

The hyperbolic area of a circle with hyperbolic radius r is $4\pi \sinh^2(\dfrac{r}{2})$. We can solve this area for its hyperbolic size and get $\ln 3$. This happens to be the only hyperbolic distance we calculated earlier; for Euclidean radius 1, the hyperbolic distance from O to a point .5 away is $\ln 3$, a distance which is easily constructed. If we construct the circle in the circle O we made using Noah's construction and hide all our marks, we get a picture of a hyperbolic circle and square with the same area.

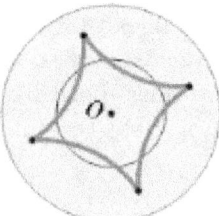

Circle and square.

2.13 Hyperbolic Trigonometry

The hyperbolic circle area formula hints at the depth of hyperbolic geometry: it has everything Euclidean geometry has, except different versions. Hyperbolic trigonometry actually uses the same hyperbolic trigonometry functions we saw in second semester calculus. We will use two hyperbolic geometry formulas which have been known for centuries. The first gives us a relationship between angles and hyperbolic lengths of sides.

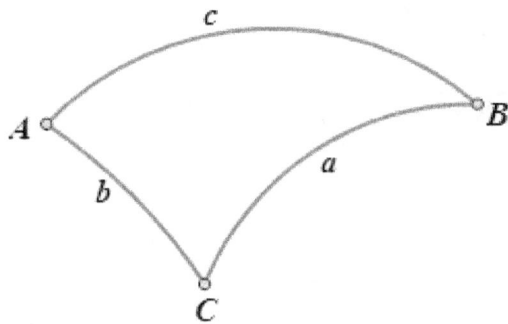

Formula 1: $\cosh c = \cosh a \cosh b - \sinh a \sinh b \cos C$

We have to keep track of that uppercase C in the cosine function because it is the only angle in the formula. Another weird property of hyperbolic space is how much its trigonometry relies on hyperbolic side lengths. But this reliance on length actually makes sense because relative lengths of the sides can contain information about location within the disk, which in turn influences just what sorts of vertex angles are possible. For instance, small triangles near the center O may have almost

Euclidean angle sums while triangles near the boundary can have angle sums close to zero. Small triangles near O will have sides with small hyperbolic lengths. The reader should be able to figure out what sorts of hyperbolic lengths occur near the boundary.

Formula 1 sort of resembles the Euclidean Law of Cosines which also uses only one angle. In fact, we could use Formula 1 to find a side length c when we know an angle C and the lengths of its two sides – just like the most obvious use of the Euclidean Law of Cosines. We will now see Formula 2 which, like the Euclidean Law of Sines, involves more angles.

Formula 2: $\sin A \sin B \cosh c = \cos C + \cos A \cos B$

Let's try an example from regular hyperbolic polygons with all right angles. We will start with an all-right angled hexagon centered at O. Suppose we join the midpoints of consecutive sides. Not only will we be able to calculate the angles and sides of the new triangles formed, we will see a rich source of hyperbolic trigonometry problems: all-right angled polygons centered at O. The central angles are easy and those right angles make sines and cosines simple.

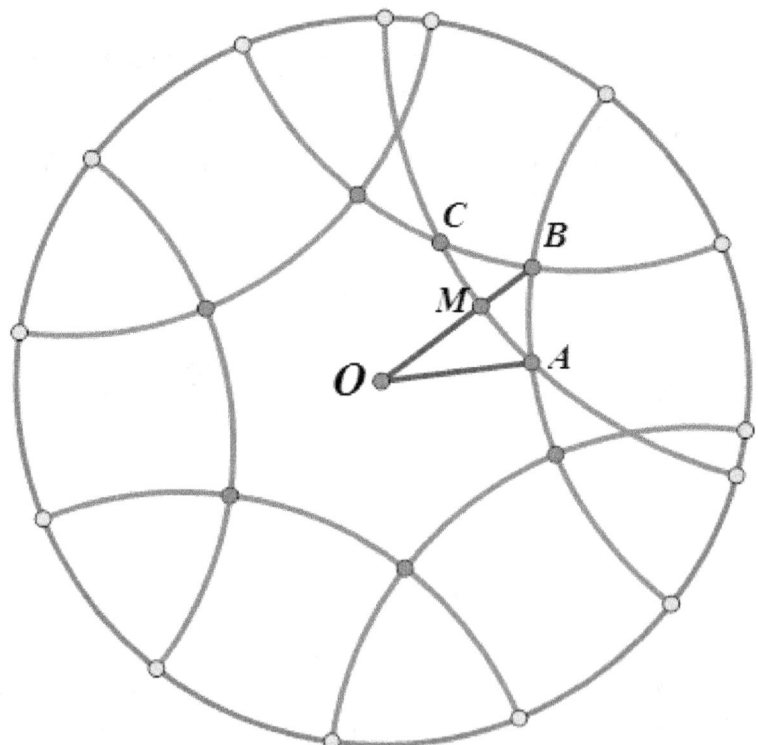

So, given the regular, all right angled hexagon pictured with side midpoints A and B, what are the hyperbolic lengths of the sides $\overline{AB}$ and $\overline{AC}$? Since we don't know any side lengths, we should use Formula 2, which requires extensive knowledge of the angles. Take a look at triangle OAB because it is a goldmine of information. Angle AOB is 30 degrees, angle OAB is 90 and ABO has to be 45 degrees (for a typically hyperbolic angle sum of 165.) Formula 2 gives us the length of any one side! Let's find all three, since it's only calculator work from here. (The steps for calculations are homework.)

$$AB = \cosh^{-1} \sqrt{1.5}, OA = \cosh^{-1} \sqrt{2}, OB = \cosh^{-1} \sqrt{3}.$$

2.14 Alhazen's billiard problem in hyperbolic geometry

The math horizon extends in both directions. Alhazen's billiard problem, for example, spans from ancient times to the present. Ptolemy first posed an astronomical question in 150 BC which Alhazen restated and explored a thousand years later. Mathematicians required another millennium to finish it off. In Euclidean geometry, Alhazen's Billiard Problem starts with a given circle and two points A and B inside. We then try to construct an inscribed, isosceles triangle with A on one leg and B on the other using compass and straightedge. The billiard aspect comes from imagining a round billiard table with two points marked on the felt. Setting up a three-cushion shot which passes through the two points and returns where it started turns out to be the same as solving the Euclidean problem.

Until Reide in1989 and Neumann in 1988, various mathematicians attempted the general construction but only obtained constructions for specific locations of the given points. Reide and Neumann each proved that the general construction could not be obtained. This information puts Alhazen's billiard problem in the non-constructible collection, along with cube root, angle trisection and other mathematical quantities unable to be obtained with compass and unmarked straightedge. Remember how squaring the circle turned out to be doable in hyperbolic geometry while impossible in Euclidean? Let's see how Alhazen's problem turns out in hyperbolic.

But before we go hyperbolic, let's get the Euclidean version details.

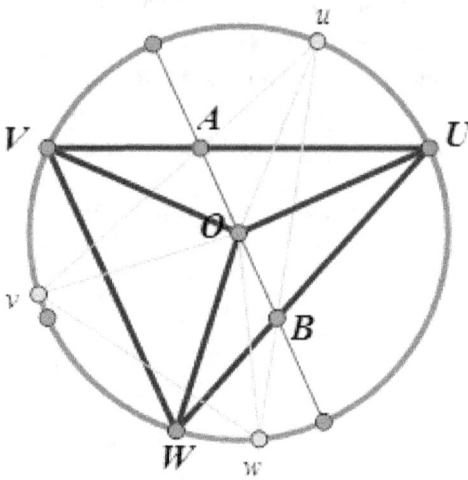

An obvious construction of an Alhazen solution exists for A and B on a diameter of the given circle and equidistant from the center, as in the figure. That's a homework problem. What's interesting is, we can quickly prove that two such triangles exist for these given points, with an extremely similar calculation for the hyperbolic case. (Depending on the location of the given points, as many as four Alhazen triangles are possible.)

The thick triangle VWU is a solution while the faint triangle designated vwu fails to be isosceles. Now we apply the Euclidean Law of Cosines to find the lengths of VU and WU and subtract.

$$VU^2 - WU^2 = 2r^2(\cos VOU - \cos WOU).$$

These central angles are always less than π when U is not an endpoint of a pictured diameter. Since cosine is decreasing for such angles, the difference is positive when $\angle VOU < \angle WOU$. The difference is continuous with value zero at the four compass points we get from the endpoints of the diameter through A and B and its perpendicular diameter.

The hyperbolic version of this same given looks similar when the hyperbolic circle has center O. The hyperbolic version has some extra structure: the boundary circle.

Here's an example.

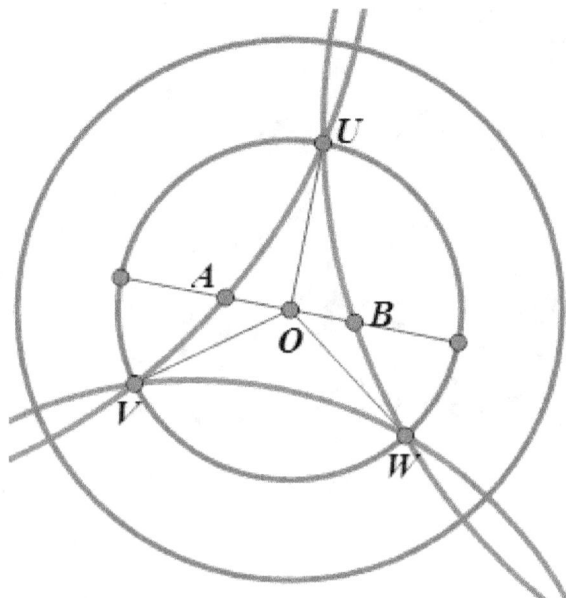

This gives us a chance to apply Formula 1 from hyperbolic trigonometry. Notice how the difference between the hyperbolic cosines of the hyperbolic lengths of the sides of the triangle turns out almost the same as the Euclidean case.

$$\cosh VU - \cosh WU = \sinh^2 r(\cos VOU - \cos WOU)$$

Since we have a subtraction of cosines of central angles, we have exactly the same argument for the number of solutions as the Euclidean case! This correspondence between the number of Euclidean and hyperbolic solutions turns out to be the case in general. Our calculations and the theorem to follow are from Nathan Poirier's 2010 summer research.

2.15 Nate's Theorem

There exists a bijection between constructible Euclidean solutions to Alhazen's Billiard Problem and the hyperbolic constructible solutions, as well as the non-constructible solutions.

Proof.   Suppose we can construct any of the hyperbolic isosceles triangles whose legs contain A and B, inscribed in a given circle with center O. We can transform, through stereographic projection, to the Klein model of hyperbolic geometry.

Stereographic projection shows how the two hyperbolic disk models are the same. A sphere rests on the center of the Klein model, with the same size radius.

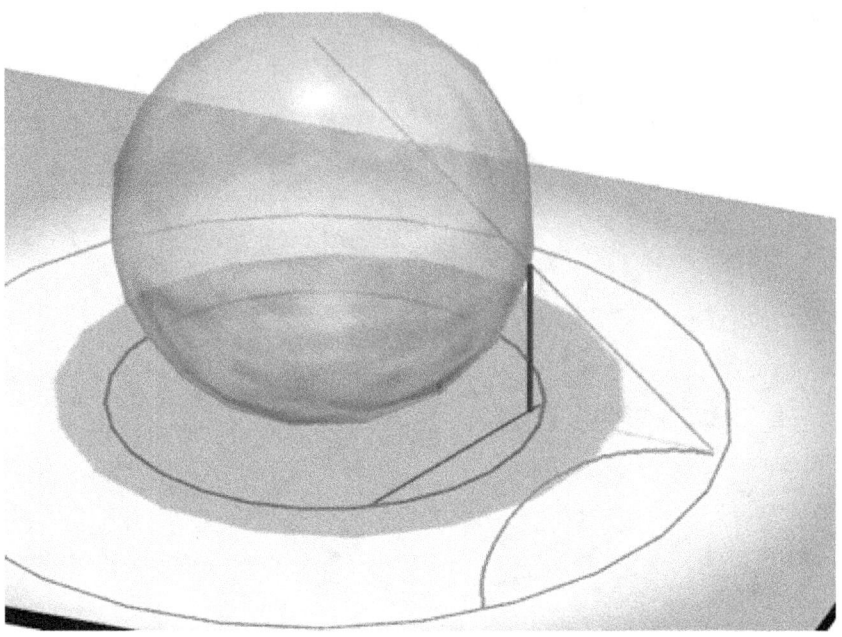

A Euclidean ray from the North Pole passes through the sphere as the ray traces a hyperbolic line on the Poincarè disk. Projecting straight down from the intersection of the ray and the sphere produces a chord in the Klein model.

Stereographic projection sends circles to circles (when they are not tangent to the boundary or not containing O) and preserves congruence and incidence. The legs of the constructed hyperbolic isosceles triangle become Euclidean segments in the Klein model. The given circle transforms to a circle. The projection can be accomplished with compass and straightedge because the vertices of the Poincaré disk triangle remain on the given circle, so we can join them with our ruler to get the Klein model triangle. (See the figure below.)The Euclidean rays $\overrightarrow{OA}$ and $\overrightarrow{OB}$ intersect the Euclidean segments at $A\prime$ and $B\prime$, the Euclidean points on the legs.

Now, when we consider this twice-transformed picture without the hyperbolic interpretation, we see an isosceles triangle with the transformed images of points A and B, one on each leg. In other words, our hyperbolic construction yields a Euclidean construction for an Alhazen solution.

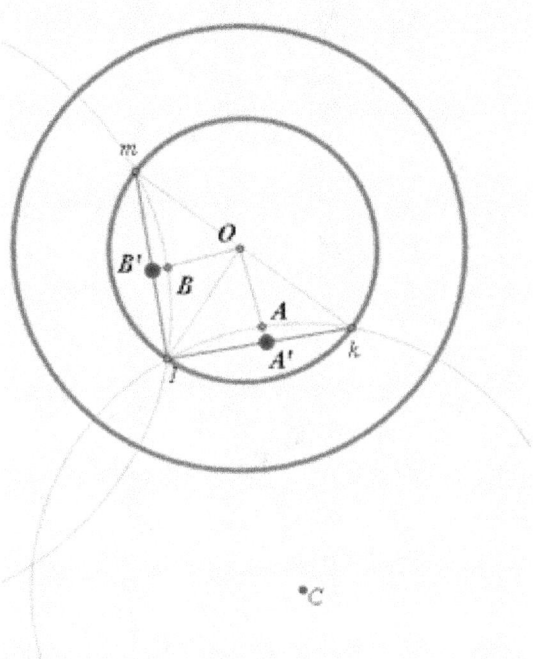

The process is reversible because stereographic projection is a bijection. If we could find any hyperbolic construction, we could then find any Euclidean construction. But, as has only recently been proved, the Euclidean solution is, in general, unobtainable. Therefore our initial supposition cannot be true. ∎

2.16 All right-angled regular hexagon example

Just like last section, we follow a long research-type problem with some examples before we get to the homework problems. Let's warm-up with a few written-out hyperbolic geometry questions. For further inspiration, remember to be flexible in your thinking. Keep the Rules handy. Remember to understand the given and try to find a way which forces the conclusion to be true from the given.

We will use the following figure for two questions. It pictures the same all-right angled regular hyperbolic hexagon that we constructed in the hyperbolic section with an all-right angled regular hyperbolic hexagon constructed using hyperbolic circles (construction marks hidden.) In other words, hexagons $ABJIHG$ and $ABCDEF$ are the same special kind of hexagon.

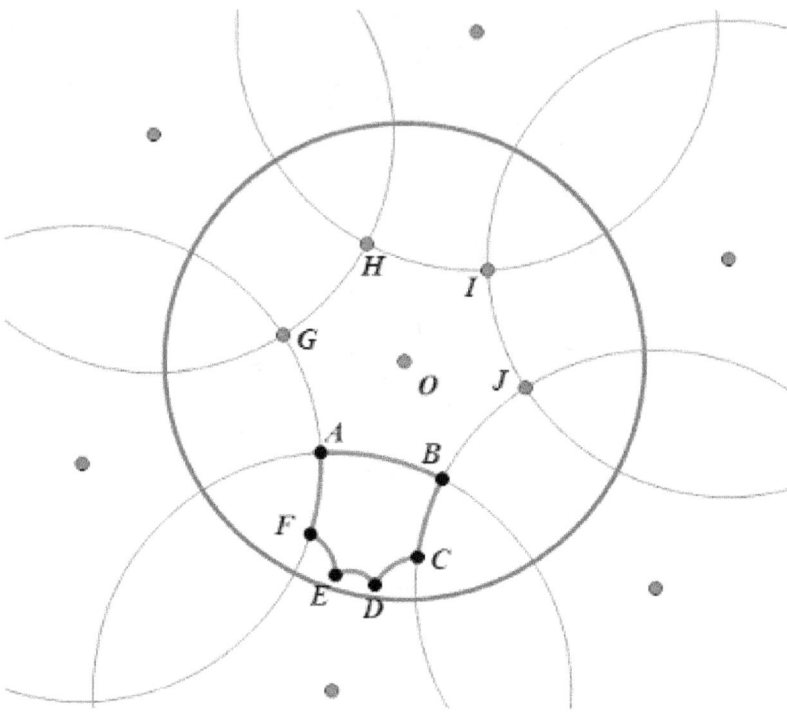

Our first question is, prove these two hexagons are hyperbolic congruent. The following answer is pretty good, but it has a small gap in the logic. Can the reader find this jump?

> Since all the angles are 90, all the angles are congruent. All the sides of each single hexagon are congruent and since they both contain side $\overline{AB}$, all the segments of both hexagons are congruent. Then the two hexagons are congruent.

The jump was that last sentence, wasn't it? The first two sentences stand as written. But that last sentence really needs a "because," since we do not have any rule, not even AAAAAA, for congruent hyperbolic hexagons.

How do we fix this jump? Let's try using isometries, placing hexagon $ABCDEF$ on hexagon $ABJIHG$ as carefully as we can and see if we reach a place where some part fails to line up. If the six vertices land on each other, we're done. If we fall short, let's see if there's enough structure to force the pieces to meet anyway.

Since both hexagons contain $\overline{AB}$, let's reflect $ABCDEF$ across $\overline{AB}$. Then sides $\overline{AF}$ and $\overline{BC}$ land on $\overline{AG}$ and $\overline{BJ}$, respectively, because the angles are all 90. Since all the sides are congruent, point F lands on G and point C lands on J. Now consider the next two pairs of corresponding sides. The segment $\overline{FE}$ lands on segment $\overline{GH}$ because they have the point F/G in common and they both have a 90 degree angle.

The side $\overline{CD}$ lands on $\overline{JI}$ for the same reasons. Since all sides on congruent, the last two pairs of points land on each other, matching all six pairs of vertices. ■

We did not hit a gap because we always had the sides and angles the same sizes. This sort of argument holds for hyperbolic polygons in general. Some writers say that there is exactly one all right-angled regular hyperbolic hexagon. This is different from Euclidean because in Euclidean we can have a regular hexagon with area as big or as small as we wish and the most we can say is all regular Euclidean hexagons are similar.

2.17 A typical logic problem

Prove that an exterior angle of a hyperbolic triangle is not equal to the sum of the remote interior angles.

Proof. Let's use a little picture to stand for any hyperbolic triangle ABC with exterior angle ACD. Notice the triangle has that pinched look common to hyperbolic triangles.

The sum of the interior angles is less than 180 degrees but the linear pair property still holds in hyperbolic geometry. So, $\angle A + \angle B + \angle ACB < 180$ and $\angle ACB + \angle ACD = 180$. Substitution and subtraction gives us $\angle A + \angle B < \angle ACD$. ■

More can be said, illustrating another type of question to be answered. How does the Euclidean theorem to which we just referred connect to the Parallel Postulate?

Since the exterior angle of a Euclidean triangle equals the sum of its remote interiors in Euclidean geometry, and the result is false in hyperbolic, the Euclidean fact must rely on the Parallel Postulate. Let's see the connection. The proof of the Euclidean result uses the sum of the angles of a triangle being 180 degrees. The proof of this angle sum uses the Parallel Postulate. (That's the proof where, for triangle ABC, we use the unique line parallel to side $\overline{BC}$ through point A and alternate interior angles to get the angle sum.)

2.18 Hyperbolic geometry topics to know

Constructions for hyperbolic lines.
  Hyperbolic distance formula.
  Construct hyperbolic midpoint of a hyperbolic segment.
  Construct hyperbolic center of hyperbolic circle.
  Sum of angles of hyperbolic triangle.
  AAA for congruent hyperbolic triangles.
  Construction for mutual perpendicular
  Construction of all right angled hyperbolic polygon.
  Hyperbolic triangle area.
  Limiting and ultra parallel lines.
  Asymptotic triangles.
  Hyperbolic perpendicular bisector.
  Hyperbolic trigonometric formulas.

  For your edification and enjoyment, we now present the

2.19 Hyperbolic Geometry Homework Problems

(1) Given two points A and B on the boundary which are not the endpoints of a diameter, construct the hyperbolic line $\overleftrightarrow{AB}$.
(2) Given point A on the boundary and point B inside the disk, not Euclidean collinear with O, construct hyperbolic line $\overleftrightarrow{AB}$.
(3) Give yourself a hyperbolic angle with vertex at O, called $\angle AOB$. Construct the hyperbolic angle bisector of $\angle AOB$.
(4) Construct a pair of congruent hyperbolic triangles AOB and COD.
(5) Calculate the hyperbolic area of a hyperbolic triangle with three angles of 30 degrees.
(6) In the figure below, prove angles CBH and HBI are supplementary.

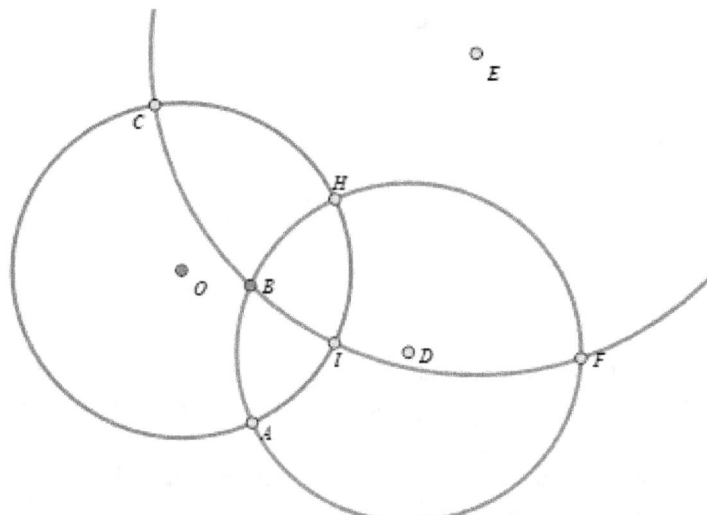

(7) Same figure as Problem **6**. Prove that hyperbolic vertical angles CBH and ABI are congruent.

(8) Same figure as Problem **6**. Prove that hyperbolic angles ABC and ABI are supplementary.

(9) Construct hyperbolic quadrilateral $OBCD$ so that angles O, B and C are each 90 degrees.

(10) Construct the mutual perpendicular for lines $\overleftrightarrow{AB}$ and $\overleftrightarrow{OC}$.

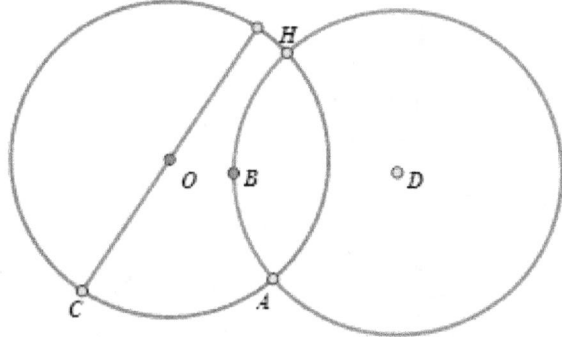

(11) Use a circle with radius 2 as the Poincaré disk. Calculate the hyperbolic distance from 0 to 1. Then do the same for a disk of radius 4 and calculate the hyperbolic distance from 0 to 2.

(12) Calculate the hyperbolic area of a hyperbolic circle with hyperbolic radius $\ln 9$.

(13) Here's a handy property to know. Given the Euclidean center C of a hyperbolic line on an extended diameter $\overleftrightarrow{OC}$, prove the hyperbolic line is perpendicular to the diameter.

(14) Given two pairs of congruent hyperbolic angles, $\angle OAB \cong \angle OCD$, and $\angle OBA \cong \angle ODC$, prove triangles ODC and OBA are congruent.

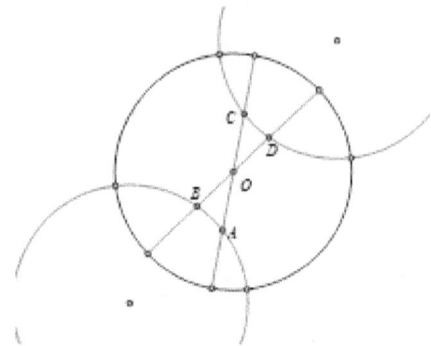

(15) Construct the unique mutual perpendicular between a diameter and a hyperbolic line whose radical axis is Euclidean parallel to the diameter. Prove your
construction works.

(16) Give yourself a scalene Euclidean triangle with an incircle and an excircle. Then
construct the circle orthogonal to both circles whose diameter has the two points
of tangency as endpoints.

(17) Given the circle hyperbolic center H, use SAS to prove the hyperbolic triangles
HCD and HBA are congruent. Also, prove they are isosceles.

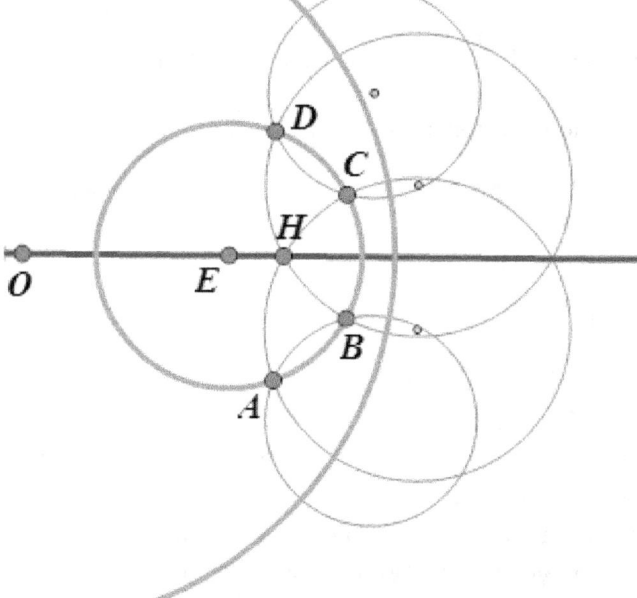

(18) Construct a hyperbolic quadrilateral whose opposite sides are ultra-parallel but
not hyperbolic congruent.

(19) What is the hyperbolic meaning of a Euclidean segment inside the disk which
does not contain the point O?

(20) Construct a pair of hyperbolic lines whose mutual perpendicular is a diameter.

(21) If a Euclidean theorem is dependent on the Parallel Postulate, what is that

theorem's status in hyperbolic geometry? Give an example of such a theorem.

(22) Parallel lines cut by a transversal have corresponding angles congruent in Euclidean geometry. What about hyperbolic geometry?

(23) Give yourself a Euclidean circle inscribed inside a Euclidean square. Then use the set-up to construct a hyperbolic quadrilateral with all asymptotic angles.

(24) In Euclidean geometry, we use the angle sum of a triangle is 180 degrees to prove the AAS method of proving triangles congruent. Does AAS hold in hyperbolic geometry? If not, construct a pair of non-congruent hyperbolic triangles (asymptotic allowed) with AAS true. If so, prove the AAS holds.

(25) Adjacent angles with exterior sides on a line are supplementary in Euclidean geometry. What about hyperbolic geometry?

(26) In hyperbolic geometry, we can prove triangles congruent with AAA. Can we prove triangles congruent with AAA in Euclidean geometry? Support your answer.

(27) In hyperbolic geometry, can we prove that through a point not on a given line, there exists more than one line parallel to the given line? Support your answer.

(28) Construct a doubly asymptotic triangle with one right angle. Calculate its area.

(29) This problem has two free-hand estimates. Just these two problems, put down your compass and straightedge and straightedge. Using only your knowledge of hyperbolic geometry, in the first drawing, mark the approximate location of the hyperbolic midpoint of hyperbolic segment $\overline{AB}$ and in the second drawing, mark the hyperbolic center H of the circle with Euclidean center E.

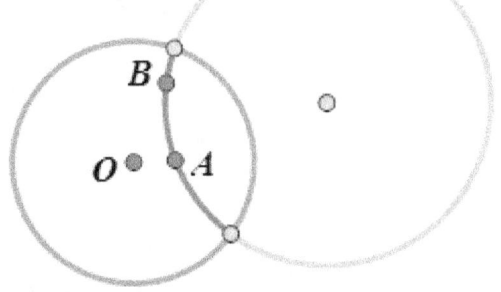 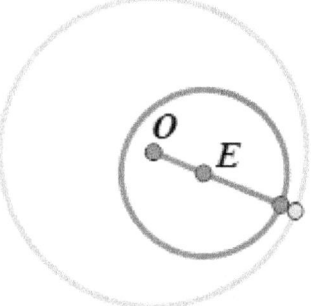

(30) When a hyperbolic circle is tangent to circle O from the inside, it is called a horocycle because it is actually not a hyperbolic circle. Why not?

(31) Where is the hyperbolic center of a horocycle? Support your answer.

(32) Calculate the $\lim\limits_{a \to 1^-} \ln(\dfrac{1 + a}{1 - a})$ and use the result to verify hyperbolic lines have infinite hyperbolic length.

(33) Prove the Euclidean construction illustrated for the mutual tangent line of two externally tangent circles in the figure captioned, "Mutual tangent."

Chapter 3

The Elliptic Interpretation of the Parallel Postulate

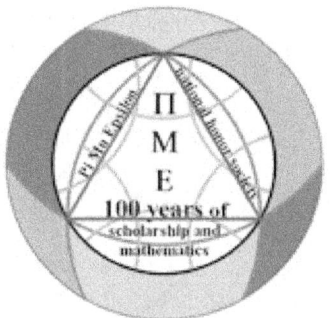

Happy Anniversary, PME!

The parallel postulate has another negation besides the one we used in hyperbolic geometry. Since the original parallel postulate mandated the existence of exactly one line parallel to a given line through a point not on the given line, we can consider the negation which mandates the existence of no parallel lines through a point not on a given line. It's only fair, really. Besides, now that we have taken the plunge and explored a non-Euclidean geometry, we might as well complete the trio. As we will soon see, the geometry of Section 3 has all the details of the other two geometries, but with its own weird differences.

Elliptic geometry "completes the set," in several crucial ways. The Euclidean plane has zero curvature. The hyperbolic paraboloid, home to one version of hyperbolic geometry, has negative curvature. The sphere, a starting place for elliptic geometry, has positive curvature. We have seen that hyperbolic triangles have angle sums less than 180 and Euclidean triangles have angle sums equal to 180 degrees. Guess what angle sums elliptic triangles have? We will soon see this and more ways that elliptic geometry characteristics, while different from the other two, fit in with them when considering the options.

3.1   The Elliptic Rules

We are jumping right into the elliptic model and saving its development for later. The elliptic points exist in and on a disk of Euclidean points whose boundary is the Euclidean circle O. We get to include the boundary points in our space this time, with the unusual proviso that points which lie on the ends of a diameter are treated as the same point. These pairs of points are called antipodal points. Diameters of circle O are elliptic lines, which is nice for us. The other form elliptic lines take are arcs of circles inside circle O which pass through antipodal points.

Constructing elliptic lines is our first order of business. Just as in the hyperbolic disk, the easiest cases involve points on the boundary. These are left for homework. Now, suppose we are given two points A and B inside circle O. There is a construction for the elliptic line through them which is almost the same as the construction from section 2, which is nice for us.

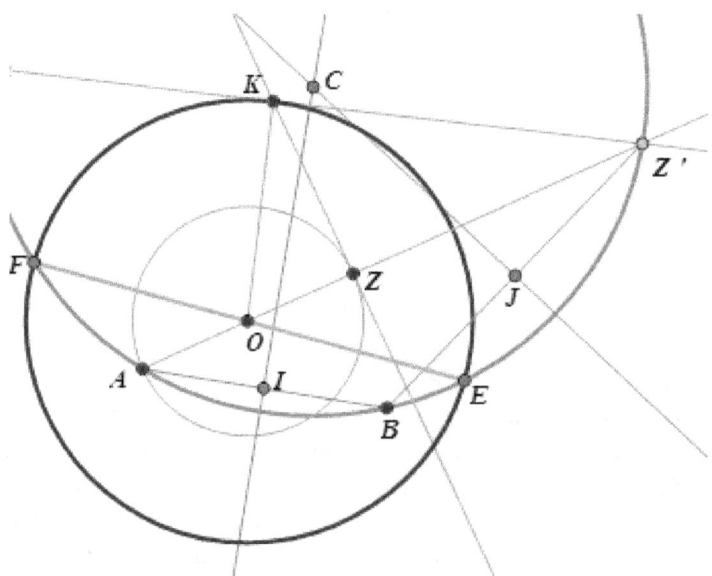

The elliptic line construction.

3.2   The general elliptic line construction

The construction starts with any two points A and B inside the disk, (besides the center O.) The first thing we do is reflect A across O. That is the purpose of the small circle containing A with center O. We call its reflection Z, as if we reflected across the alphabet. Next, we construct the inverse of Z across circle O, labeled $Z\prime$. We will prove that the circle through A, B, and $Z\prime$ is an elliptic line. For the construction, we only need the perpendicular bisectors of two segments using pairs of these points as endpoints. In the figure, we used segments $\overline{AB}$ and $\overline{BZ\prime}$. We could have reflected point B to start with, if it had been in a more convenient spot than A. We could have used the perpendicular bisector of $AZ\prime$ as well. These options are for our convenience when we have to do the actual constructions.

At first glance, the elliptic line construction might look more complicated than the hyperbolic construction. Actually, the elliptic case is exactly the same amount of work, except for the reflection of point A. What is actually remarkable is that the construction of the elliptic line is not only the same amount of work, but also the same steps as the hyperbolic line construction, once the reflection has been done.

To prove the construction works, we are given two points A and B inside the elliptic disk, we will prove that the circle determined by the inverse of the reflection of one point and the two given points goes through a pair of antipodal points of the disk. We will also prove that a Euclidean line through O and a point B on the elliptic line intersects the circle containing the elliptic line at the inverse of the reflection of B.

Proof: We are cheating a bit in the figure because we have labeled the inverses of reflections of points both A and B when we have yet to prove that inverses of reflected points actually do land on the elliptic line. First, let's note that the three points A, B, and $Z\prime$ determine their circle. The point B could have been anywhere, so we will let it stand for any point on the circle. We will now prove that $Y\prime$ actually is the inverse of the reflection of B, as its name implies. Since the inscribed angles are pair-wise congruent, we get $\triangle OY\prime Z\prime \sim \triangle OAB$. Then we can use the proportion of corresponding sides to get

$\dfrac{OZ\prime}{OB} = \dfrac{OY\prime}{OA}$ which gives us $OZ\prime \times OA = OB \times OY\prime$. Substituting two equal lengths gives us $OZ\prime \times OZ = OY \times OY\prime$.

The last equality comes from our reflections across O to get Y and Z. The last equality proves that the point we labeled $Y\prime$ is in fact the inverse of Y. This is important because it means that any Euclidean line through O which intersects our constructed circle twice gives us a pair of points which are inverted reflections of each other. This property is the key to showing that the circle we have constructed passes through a pair of antipodal points.

3.3   The elliptic line construction proved

In our figure, there is a segment $\overline{EF}$ which appears to be a diameter. Again, we are cheating a bit in our drawing because we have drawn what we are going to prove. Let E be a place where our prospective elliptic line intersects the boundary circle.

Now, the Euclidean ray $\overrightarrow{EO}$ intersects that prospective elliptic line at a point we call F, careful not assume F is on the boundary circle, even though such appears to be the case. This point F has to be the inverse of the reflection of E across O. Well, the reflection across O of E would be the antipodal point of E. The inverse of the antipodal of E would remain right where it is because this point is on the boundary! (As we have seen in a homework problem at the end of Section 1, the inverse of a point on the circle of inversion is the point itself.) This means the point F is the antipodal of E, which means the circle we constructed meets the qualification for being an elliptic line.   ■

Welcome to our new playground. Our first exploration will be to see how this interpretation of points and lines satisfies the axioms.

The first axiom says that two points determine a line. This is the reason for the identification of antipodal points. We can construct lots of elliptic lines through one fixed pair of antipodal points, which is a homework problem. So designating two antipodal points is really like naming only one point. As in Euclidean geometry, we can construct lots of lines through one point. But as soon as we have two non-antipodal points designated, there is exactly one line through them.

The second axiom says that an elliptic line can be extended indefinitely. The identification of antipodals satisfies this requirement as well because we can travel an elliptic line indefinitely, like going around in a circle. Actually, a bug traveling north on an elliptic line simply reappears at the line's southernmost antipodal point as soon as the bug hits the northern antipodal point. The bug just keeps going, never leaving the line, never hitting an elliptic endpoint.

The third axiom says we have circles. As in the hyperbolic case, the elliptic center is not the Euclidean center of the circle. We will see how this works out later.

The fourth axiom requires all the right angles to be congruent. In this disk, we use tangents to measure angles again, so even curvy-looking right angles have the same angle size as two perpendicular diameters.

Our elliptic negation of the fifth axiom says that through a point not on a given line, there is no line parallel to the given line. In short, we have no parallel lines in this space. Any elliptic line divides the disk into two sides. For any point not on the line, it is on one side of the given line. Any elliptic line through this point has to contain a pair of antipodals, so it will have to contain points from both sides of the given line. So there is no way to avoid the given line, our new line will have to cross it.

Our brief history of non-Euclidean geometry implies that hyperbolic geometry was the first non-Euclidean geometry which mathematicians explored. This is inaccurate. The shape of the earth forced explorers to find ways to navigate the trackless oceans. The sun in the sky had to be used because for half of each day, it was the only visual reference. The earth's rotation and the time of year had to be taken into account, as well. For instance, the sun is not necessarily directly overhead just because it is noon. The complicated processes which led to the inventions of dependable time pieces and sextants made some people wealthy and left others to waste away, lost at sea.

The shortest distance between two points on a globe is not a Euclidean line. This is easy to see because the Euclidean line through Passaic, NJ and Paris, France goes through the earth, an unavailable route. A good way to see the shortest route on the surface would be to find a globe, a friend and a shoelace. Have the friend pinch the shoelace at Paris and Passaic, while the geometer gently pulls the shoelace tight. The shoelace forms an arc which is part of a great circle, that is, a circle with the center of the sphere as its center. Yet another way to see these kinds of arcs on earth is to peek at the flight paths of aircraft printed on a rectangular map. The flight paths appear curved when, in fact, the paths are straight on a round world. When the map of the world is flattened for a paper map, the world is distorted and the paths appear curved.

3.4   Elliptic area

The sphere is actually the source for our elliptic disk model. Students might find it useful to think of our disk as a flattened hemisphere, with those antipodals paired up. Remember that we can construct two elliptic lines through a pair of antipodals and these lines enclose a region. This region, on a sphere, is called a lune. It is easy to calculate the area of a lune and this is left as a homework problem. If the angle formed by the two lines is α radians, then the area of the lune is $2\alpha r^2$, where r is the radius of the sphere. We will use this formula to find the area of an elliptic triangle.

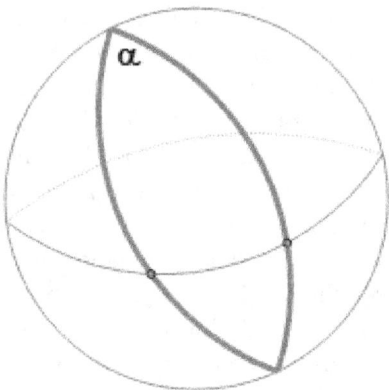

Lune on a globe.

The figure below is a clear globe with three great circles on it. The dashed parts are the arcs of great circles on the other side of the globe. The triangle at the top has its vertices labeled. The pale triangle near the bottom is meant to be the twin triangle formed on the other side of the world by these great circles.

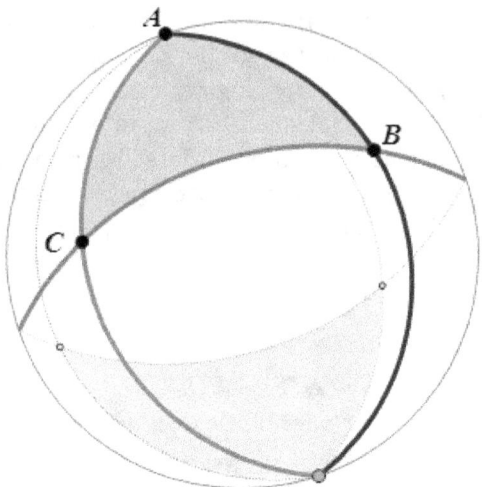

Three great circles on a globe.

Let the radian measures of the angle in $\triangle ABC$ be α, β, and γ, respectively. We will now prove that the area of an elliptic triangle is $(\alpha + \beta + \gamma - \pi)r^2$.

Proof: The lune facing us formed by the great circles through A has a twin lune on the other side, bounded by the dashed arcs. If we let the angle between them be called α (to correspond with the point A), we know the area inside the lune is $2\alpha r^2$. Similarly, the area inside the lune bounded by the circles through B is $2\beta r^2$ and the area inside the lune bounded by the circles through C is $2\gamma r^2$. We have to remember that each of these lunes has a twin on the other side of the globe. If we add up all the lunes with their twins, the entire globe is covered and our two shaded triangles are each covered three times.

Let's be careful here. We already know a formula for the area of a sphere, $4\pi r^2$. If we add up all the lunes and their twins and then subtract the extra coverings of the triangles, we could set the calculation equal to the area of a sphere. We don't know the area of the triangle, so we'll use a variable for it and solve for that variable, x.

$$2(2\alpha r^2 + 2\beta r^2 + 2\gamma r^2) - 4x = 4\pi r^2$$

Solving for x gives us the formula for area. ■

It is worth summarizing the situation. If we treat the constant radius in hyperbolic and elliptic geometries as 1, hyperbolic triangles have areas between 0 and π, not inclusive. Their angle sums have the same limitations. Euclidean triangles can have any finite area and their angle sums are always π. Elliptic triangles have areas limited by the angles we can get in a hemisphere because the elliptic model does not use the whole globe. The biggest triangle approaches the entire disk, with each side approaching the boundary and each angle approaching π. So elliptic triangles have an upper limit on their areas of 2π. Their angle sum has lower limit π and upper limit 3π.

The angle sums of triangles in the three geometries fit together, 0 to π; π; π to 3π. We will find other properties of these geometries which mesh. Such facts provide more reason to group these three geometries together. As we explore elliptic geometry, we will find properties that work only in the non-Euclidean geometries. Being able to adjust the sums of the angles of polygons will prove to be a key in these non-Euclidean specialties. Although mechanical engineers rely on the angle sum of a triangle being fixed at 180 degrees, we explorers will find that the freedom to play with the angle sums of triangles opens up some truly amazing constructions. We could say that one of the signal satisfactions of non-Euclidean geometry is our ability to do things which were impossible in Euclidean geometry.

3.5   Elliptic triangle angle sum and midpoint

We just found the sum of the angles of an elliptic triangle from the area formula for a sphere. We could have deduced the same limitations by considering a triangle in the disk. Here we have repeated the set-up from the hyperbolic disk: translate a generic triangle so that a vertex lands on O. Then the tangents to the other two vertices are drawn. It is worth comparing how the intersection of the tangents compares with the hyperbolic case.

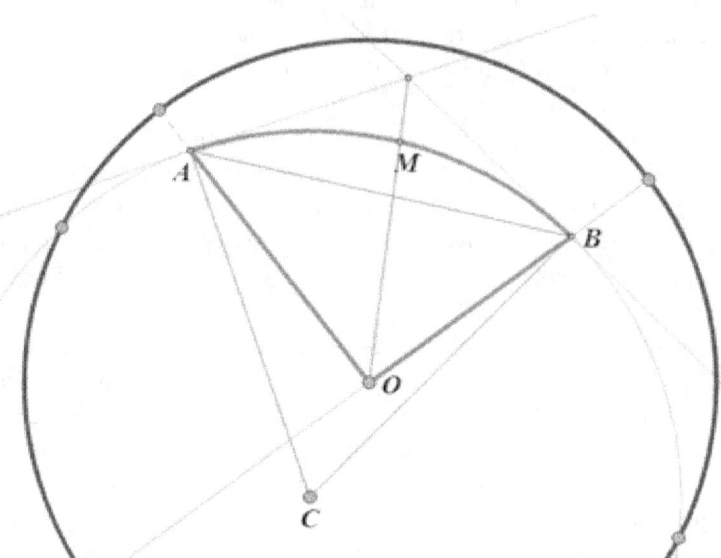

Elliptic triangle and Euclidean tangents.

Those tangents meeting outside the Euclidean segment $\overline{AB}$ implies the angle sum must exceed the Euclidean sum. Letting the angles all approach π gives us our upper bound.

You can expect questions which rely on knowledge of these bounds. Just as in hyperbolic geometry we could not have Euclidean-sized angle sums, neither can we have them here. Just as in hyperbolic geometry new objects were possible, (like a hexagon with all right angles,) new objects are available here in elliptic space.

We skipped the construction of the elliptic center of a circle in the elliptic disk. The same ideas we used in the hyperbolic case apply: we need an elliptic line orthogonal to the elliptic circle and the Euclidean line through the circle's center and O. With our vision of the elliptic disk as a squashed hemisphere, we can see the elliptic center compared with the Euclidean center. Imagine a hemisphere wearing a skullcap. The bead marking the center of the skullcap is the elliptic center of the circle bounding the skull cap. Inside the hemisphere is the Euclidean center of the bounding circle. When we squish the hemisphere into the disk, these points don't necessarily land on each other. But, unlike the hyperbolic case, they don't get far from each other, either.

The drawing for the elliptic triangle above is doing double duty: same drawing for the elliptic midpoint. If we pretend we started with the little circle, then M is on the diameter of an undrawn circle and on the diameter through O. Therefore, M is the elliptic midpoint of the elliptic segment $\overline{AB}$. The undrawn Euclidean midpoint of arc AB is not far from the elliptic midpoint. The drawing also illustrates the construction of the elliptic midpoint of an elliptic segment. The same moves from the hyperbolic case apply again.

3.6 Poles and Polars

We had good luck exploring the polygons with all right angles in hyperbolic geometry, so we will tackle the same family here. A quick calculation (saved for a future problem) shows us the family is a small one: elliptic triangles and one other case we have seen are the only elliptic polygons which allow all right angles. Such triangles are easy to see on a globe. We start at the equator and draw to the North Pole. Then we make a ninety degree turn and head back to the equator. To close the triangle, head for the starting point by the shorter route along the Equator and we have drawn a triangle with all right angles.

There is a relationship between the North Pole and the equator which did not happen in the previous sections: any great circle orthogonal to the equator passes through the North Pole. The geometric vocabulary for this arrangement is pole and polar. In elliptic geometry, every elliptic line has a pole: the *pole* is the point through which all elliptic lines perpendicular to the given line must pass. The given elliptic line is the *polar* for that point. The reader is asked to think about how we would construct the pole for a given elliptic line.

We are given an elliptic line which we will treat as a polar and we seek its pole. This is the point which acts like the North Pole when the polar is the equator. All elliptic lines perpendicular to the polar pass through this pole. Our first construction for the pole relies on common sense: if we build two elliptic lines perpendicular to the polar, their intersection will be the pole. As long as the given polar is not a diameter, there is one elliptic line perpendicular to the polar which we can find very easily, using only the straightedge.

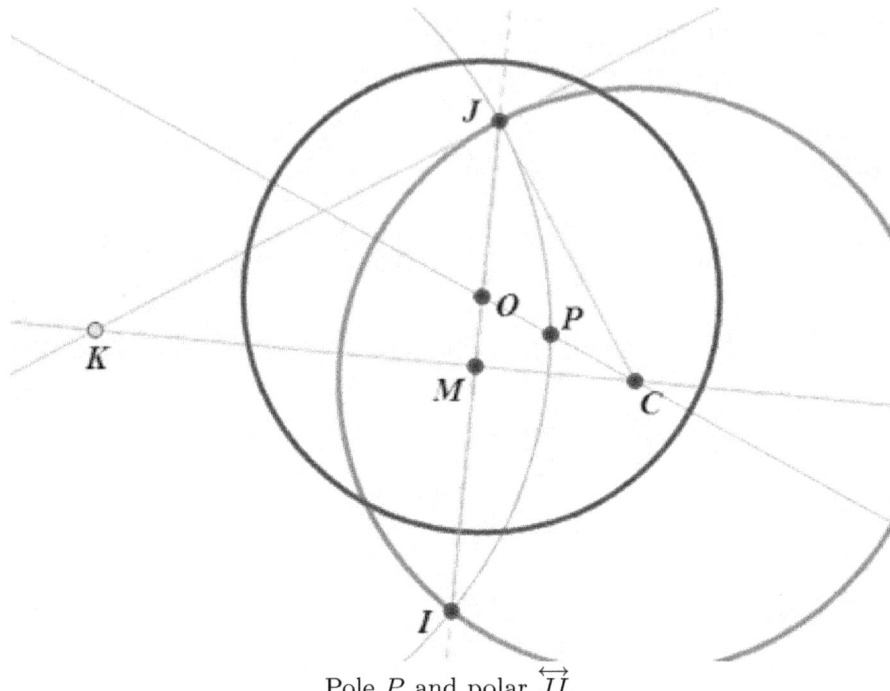

Pole P and polar $\overleftrightarrow{JI}$.

We can lay the straightedge across O and the center C of the given elliptic line and draw the diameter perpendicular to the given elliptic line, which is drawn thin in the figure above. Then we get to pick any point we want on the polar, J in the figure above. We construct the tangent at J. Then we lay down the straightedge across J and O to find the inverse of the reflection of J, the point I in the figure above. Next, we construct the perpendicular bisector of the Euclidean segment $\overline{JI}$.

The perpendicular bisector and the tangent intersect at the center of the elliptic line we want, the point K in the figure. The circle with center at K and radius $\overline{KJ}$ has to be an elliptic circle because it passes through a point and its inverse reflection. We built its radius perpendicular to a radius of the polar. Therefore the new line must be perpendicular to the polar. This new line intersects the line through the given center and O at the point P, which is the pole.

Poles and polars give us a way to view how certain lines and points are related in elliptic geometry. Every elliptic line has its pole and their construction can be tricky because of the antipodal points.

Even though that was a fairly quick construction, we will see how to save a few steps in the next drawing. In fact, there are many ways to construct poles. The faster construction of the pole uses a clever choice for the point J in the previous construction: an antipodal point we call A.

The given elliptic line through A is given with Euclidean center C. The Euclidean line $\overleftrightarrow{OC}$ through the centers gets drawn first. Then we draw Euclidean segment $\overline{AC}$, then the Euclidean segment perpendicular to $\overline{AC}$ at A. This new segment intersects the (possibly extended) diameter $\overline{OC}$ at point B. We put our compass point at B, the pencil at A and draw an elliptic line perpendicular to the given line which passes through the diameter at the pole P. Done.

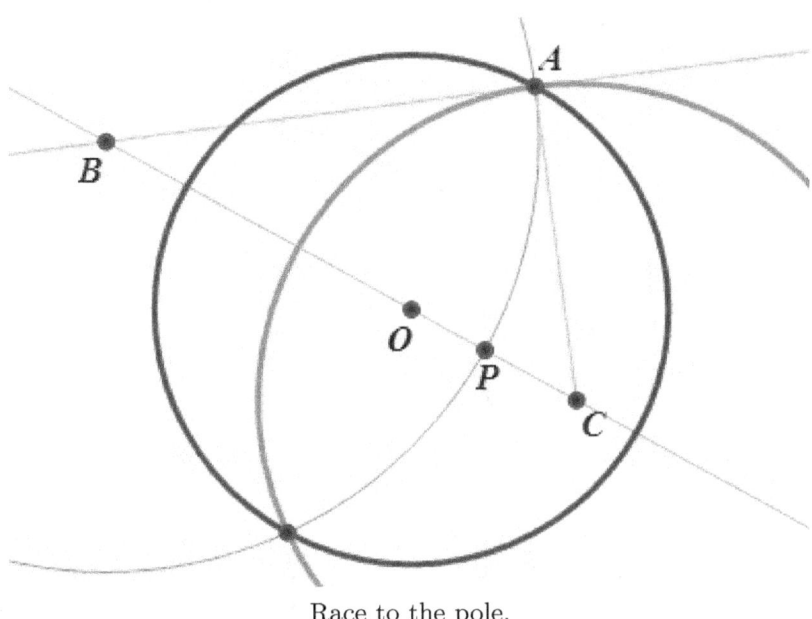

Race to the pole.

We just found a pole for a given polar. In the homework, we turn it around ask for a polar given a pole. It is true that all the poles and polars have to be in the elliptic disk, which means we have an invisible structure in the elliptic disk pairing points and lines. Every elliptic point has a unique polar and every elliptic line has a unique pole. Be especially prepared to handle some particular cases of poles and polars for special situations.

3.7 All right-angled triangles

In the figure below, we have an elliptic triangle ABC with all right angles and no construction marks. The homework requests its construction. Here's a hint: elliptic line $\overleftrightarrow{AB}$ is polar to pole C. Much more can be said... in the homework. Look for that far easier examples of 90/90/90 triangles there.

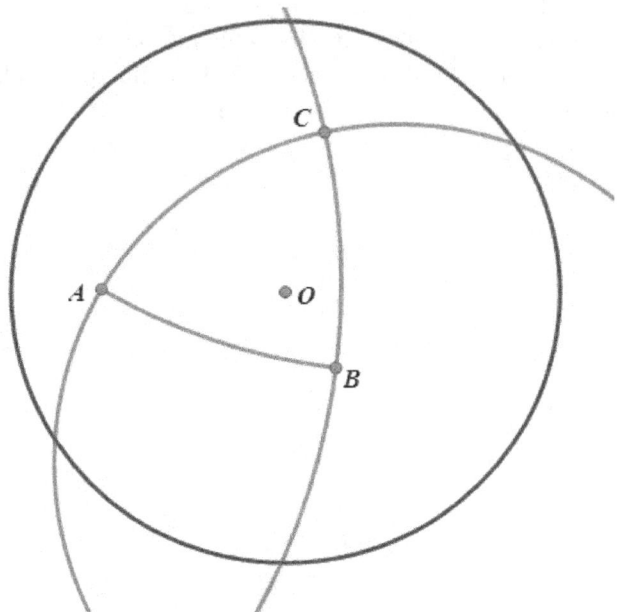

All right angled triangle.

We should be able to apply pole and polar strategy to any two perpendicular elliptic lines. We can construct an example of a triangle which would be impossible in Euclidean geometry: an isosceles elliptic triangle whose base angles are right angles. We would have to construct the first two elliptic lines with the same Euclidean radius. Then their mutual perpendicular would be symmetric to each line, giving us the isosceles property without worrying about the elliptic distance formula. Just like the hyperbolic disk, the center of our elliptic space is the only point where the elliptic center of a circle would be the Euclidean center. So, when we can construct congruent Euclidean objects symmetric around O, we have objects which are elliptically congruent.

3.8   Elliptic distance

For the sake of navigating on a globe, we will take a look at a formula for calculating elliptic distance. It is the spherical distance formula found in some secondary textbooks. We will take advantage of our focus on constructions to give the formula without proof. We recall that the hyperbolic distance formula referenced complex points on the Euclidean ends of hyperbolic lines. The spherical distance formula references a north pole, N. It also refers to angles we can measure using latitude and longitude, and that is how we will work the example. We will find the distance between points A and B on the Earth. The formula is:

$$\cos n = \cos a \cos b + \sin a \sin b \cos N$$

Notice all the trig functions are Euclidean trig functions. That means n, a, b, are arcs on the surface of the Earth and N is an angle between arcs. We will use a bit of a simplification: the lower case arcs match angles measured from the center of the Earth. Let's use the formula.

Let point A have position 40°W, 50°N and B have 10°W, 20° N. To find the distance between A and B, we need the angle from the center of the Earth which cuts arc AB. That angle is n. If you have difficulty seeing how the latitude and longitude numbers have changed between the given information and the formula, draw two circles. On one circle, draw the earth viewed from above the North Pole: you will see where the 30 comes from. On the other circle, draw the earth viewed from the side with point A on the circle itself, and mark the Equator: you will see where the 40 comes from. Then you can figure out where the 70 comes from.

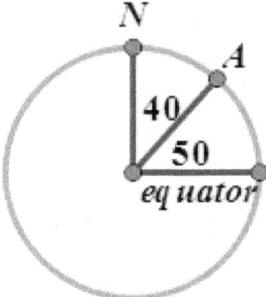

$\cos n = \cos 70 \cos 40 + \sin 70 \sin 40 \cos 30 = .7851$.

Now we need n in radians because arc length is radius times radians. So, the distance between A and B on the surface of the Earth is .668 times 3960 miles, about 2645 miles.

3.9   Reuleaux triangles

If we take a small digression to the Reuleaux triangle, we will see one of the easiest elliptic line constructions. We will treat the digression as another research example. This digression will also explain some of the structure of the Pi Mu Epsilon anniversary button design used at the start of this section.

    The Reuleaux triangle is easy to construct. From a set segment, we construct an equilateral triangle. Then we place the point of the compass at vertex and use the length of a side as a radius. An arc is drawn outside each side, with the arcs meeting at the vertices of the triangle.

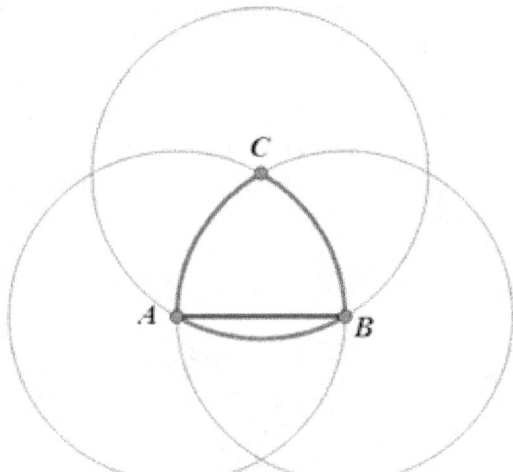

Reuleaux triangle ABC.

    Two of the circles we use to construct the equilateral triangle contain the arcs of two of the sides of the thickened Reuleaux triangle. This object has constant width, but it is not a circle. To see the constant width, note that the longest segment inside the Reuleaux triangle from any point is to the opposite vertex. Its construction indicates that this length is always the same. *Mathworld* has an excellent summary of the properties of this object. But a careful search of pages regarding its properties will not reveal the truth we are about to prove.

We will begin our research in a natural way. What easy elliptic line construction can we get from a drawing like the above? For some circle around the Reuleaux triangle, the three circles could contain elliptic lines. The circumcircle is too small: the arcs inside obviously miss antipodal pairs. The reader is urged to consider a size which looks like the circles contain antipodal pairs.

Here are some suggestions.

(1) The circle which makes the three circles contain elliptic lines is the 9-pt circle of the given triangle.
(2) The circle we seek has radius the same size as segment $\overline{AB}$.
(3) The circle we seek is the incircle of a square made of 4 copies of segment $\overline{AB}$.
(4) The circle we seek has twice the area of the circumcircle of the given triangle.

The reader is invited to mull this one over.

The fourth choice turns out to be true. There's a lot of good geometry to explore in this little problem. Reuleaux triangles as Euclidean objects have physical applications as manhole covers, as pistons in rotary engines and as drill-bits which cut square holes. A Reuleaux triangle has constant width, which means that the longest internal segment which can be drawn from any point on the triangle is constant. Circles obviously have this property. If only the triangle had an easier name to spell! The figure below shows a cast aluminum base for an outdoor lamp. Not only can we see two Reuleaux triangles in the photo; the reader should come back to this image after the next section and imagine the ciruclar arcs forming the sides of the larger triangle extended to the boundary. The result should look familiar.

Base for lamp.

The figure below has the crucial stuff drawn in. The two concentric circles have radius sizes r, the radius of the circumcircle, and R, the radius of the circle we construct to have twice the area of the circumcircle. The last homework section includes properties of this figure. For now, we'll prove that the circles forming the sides of a Reuleaux triangle contain elliptic lines when we use the circle with twice the area of the circumcircle, concentric with the circumcircle, as the elliptic disk.

3.10   Easy elliptic lines from the Reuleaux triangle

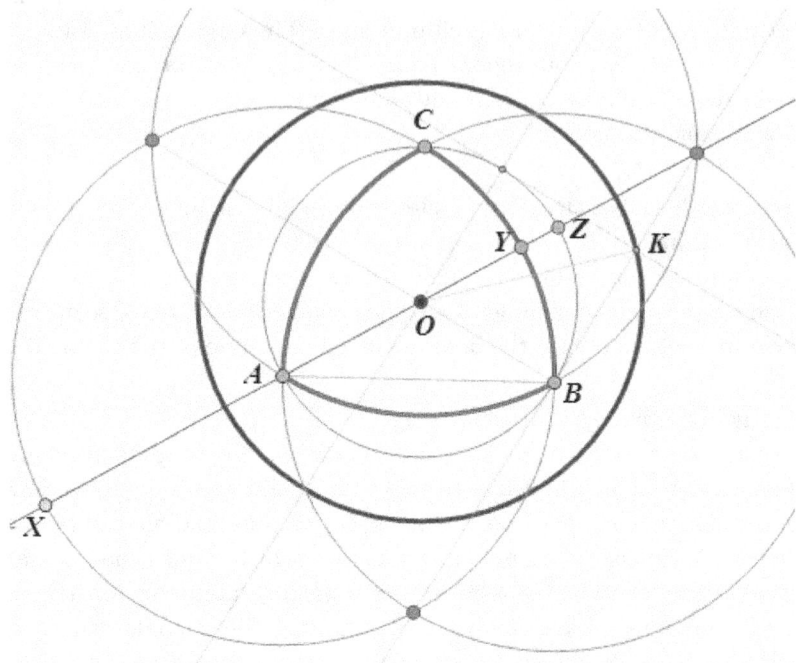

Proof: We want to prove that the thin circle containing one side of the Reuleaux triangle is an elliptic line. We're going to use the property we proved earlier, that a circle through a point and its reflection's inverse is an elliptic line. This means we want the following equation:

$$OX \times OY = R^2 = 2r^2.$$

That our construction gives $R^2 = 2r^2$ is a homework problem and it's a plain old Euclidean question. Our construction gives us the following values for OX and OY.

$OX = AX + OA = AY + r$

$OY = AY - OA = AY - r$

The reader is strongly urged to put it all together now. It's okay to peek at the homework questions for further hints. We're certainly very close to our desired equation. Here we go.

$OX \times OY = (AY + r)(AY - r) = AY^2 - r^2 = (\sqrt{3}r)^2 - r^2 = 2r^2$ ∎

3.11   Squaring the elliptic circle: a research example

The ancient Greeks could not construct a square and circle with the same area using compass and straight edge. Mathematicians for two thousand years after them could not square the circle either. In 1832, Janos Bolyai gave a strategy for constructing a hyperbolic square and circle with equal hyperbolic area. Fifty years later, Lindemann proved that π is not algebraic, but transcendental. Since the constructible numbers are not transcendental, his work implied that squaring the circle is impossible in Euclidean geometry. Bolyai's work gave an unexpected twist to this ancient area of study: before Lindemann shut down the Euclidean options, Bolyai opened a new door into the hyperbolic reality.

In the hyperbolic section, we saw Noah Davis's construction for squaring the circle in Poincaré's hyperbolic disk. Noah showed that there were countably infinite many ways to construct a hyperbolic regular quadrilateral with the same hyperbolic area as a hyperbolic circle which was also constructible.

This leaves one more geometry to consider: elliptic. In these next few pages, we see how this ancient story ends. We will follow the work of Kyle Jansens and Noah Davis, who found a construction for squaring the circle in the elliptic disk and calculated the Euclidean size radius necessary for a circle to match the elliptic area. As tricky as it sounds, the calculations have more unexpected tricks.

Elliptic geometry has its own trigonometry, of which we have only seen one formula, the distance formula we used on the Earth. In order to square the circle in elliptic geometry, we need the elliptic area formula for an elliptic circle: Area $= 2\pi(1 - \cos r)$. Not only can the interested reader find all the trig formulas in books and on the internet; but this area formula can be done as a surface integral on a globe of radius 1. That little r is the elliptic radius of an elliptic circle, not the radius of the globe or the disk.

The only way to square the circle in non-Euclidean geometry is to perform calculations ahead of time. We have to make sure the distance and angle we need are constructible and we have to calculate what Euclidean sizes we start with so that all the elliptic pieces will be the right size. Here we go.

If we choose elliptic radius $r = \dfrac{\pi}{3}$, good things happen. That area formula gives us elliptic area π and we calculate our vertex angles of the elliptic square have to be $\dfrac{3\pi}{4}$. The construction of the elliptic square will start with angle of $\dfrac{\pi}{8}$, for reasons will see when we do the construction. We will save the calculation of the Euclidean radius of the elliptic circle for last. The construction will demonstrate that we find the elliptic disk at the end of the elliptic square construction and we cannot find the elliptic radius until we have the disk.

The figure below is the start of the construction. There are tick marks on a visible coordinate system because that made it easy to get started.

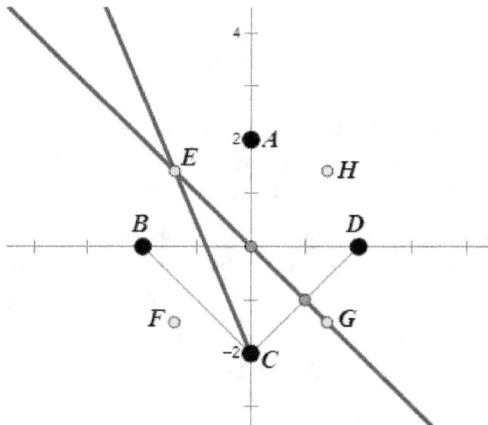

Elliptic square start.

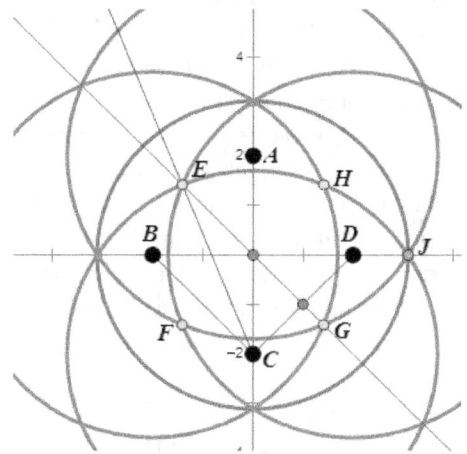

Elliptic square finished.

The Euclidean centers will be A, B, C, and D. We will make four circles, with radii AF, DF and radii BH, CH.

The rest of the construction can be summarized quickly. The four circles will be our elliptic lines containing the sides of our elliptic square. When we see where the prospective elliptic lines cross each other beyond the vertices of the square, we will know what size to make our disk so that the circles end up being elliptic lines, radius $\overline{OJ}$.

Constructing the disk last is a nifty trick indeed! We can do all the fancy stuff first and then find the disk to make all the Euclidean circles into elliptic circles.

In the figure above, it is easy to construct the tangents at point F and use them to calculate the vertex angle of the square $EFGH$, which turns out to be $\frac{3\pi}{4}$, the size we needed.

One more task awaits us: construct the elliptic circle with center O and elliptic area π. The area formula for the elliptic circle says that the elliptic radius needs to be $\frac{\pi}{3}$. Calculating the Euclidean distance requires another spherical trigonometry formula and more geometry (see their article and our homework) and we get $x = \frac{1}{\sqrt{3}}$, a constructible length. With a little planning ahead of time, we can construct this distance in our disk.

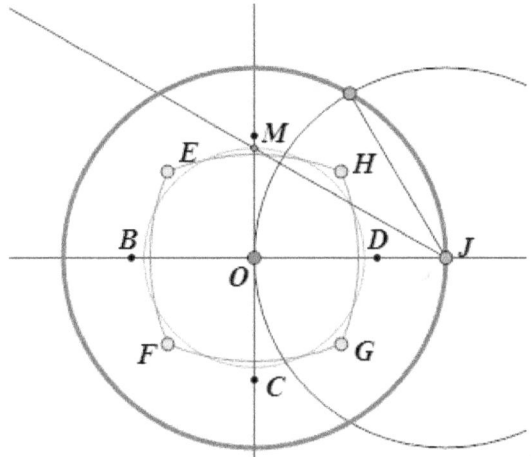

Elliptic circle and square.

The thin circle with radius $\overline{OM}$ has been constructed using angle MJO of 30 degrees. (We treat $\overline{OJ}$ as the unit length.) We can see the square and circle overlapping each other in the only way in which the two objects could have the same area.

There is one more interesting characteristic of squaring the circle in non-Euclidean geometries: the square and circle must be constructed separately, after calculations tell what Euclidean sizes to use for the starter angle in the square construction and for the radius in the circle construction. The hyperbolic version of this was first proved in 1996 by Jagy and he proved it in Bolyai's original context, without the Poincaré disk model or any other model. If the reader has done many constructions in using this book, this result rings particularly weird because we have almost always built on the given. Jagy for the hyperbolic and Kyle and Noah for the elliptic tell us that, even if we were given one of the square or circle, we would have to build the other object from measurements of the given object, not from the object itself. We would have to perform calculations in any event to square the circle in non-Euclidean geometry. This last bit of research just goes to show how strange and satisfying the open-ended geometry problems can be.

3.12 Elliptic examples

Given a lune formed from two elliptic lines whose Euclidean centers are endpoints of a diameter of circle O, prove the area of the lune is π.

Let's start with a quick drawing and look for useful structure.

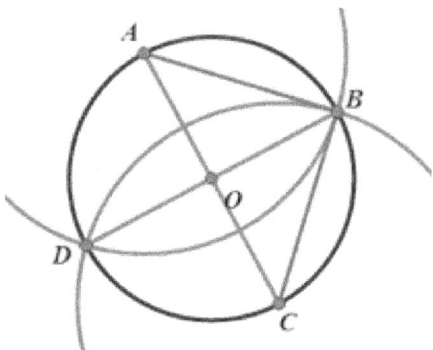

The Euclidean angle ABC has to be 90 degrees because the angle is inscribed in a semicircle. Since the area formula for a lune is twice the angle (times the radius, which we take as 1), we are finished. Here's a good answer, using the figure.

> The radii $\overline{AB}$ and $\overline{CB}$ meet at a right angle because Euclidean angle ABC is inscribed in a semicircle. Then these segments are the tangents to circles A and C. Thus the lune with vertex angle at B with point O inside has area $2(\pi/2) = \pi$. ∎

Notice how we got the measure of the angle of the lune from its tangents? That was an extremely cooperative problem because the tangents were so easy to find.

Another strategy would have been to work backwards. Knowing the lune's area was π implies the lune's angle had to be $\dfrac{\pi}{2}$. Once we have the drawing, we could look for a reason that the elliptic angle at B has to measure $\dfrac{\pi}{2}$. The inscribed angle idea might just have popped up then.

Elliptic triangle

Construct a Euclidean equilateral triangle with vertices O, A and B, all inside the elliptic disk. Then construct the elliptic triangle with the same vertices.

This looks to be one of those problems where we just do what the problem says to do. Since we can construct an elliptic line through any two points, this should not be too tough an assignment.

The following figure is meant to show the given, some work and the final answer. If the reader can follow those three drawings, the reader has learned some elliptic geometry for sure. The first step is Euclidean equilateral triangle OAB. The middle drawing shows the construction of the inverse of the reflection of A, preparation for finding the elliptic line $\overleftrightarrow{AB}$. The last drawing constructs the Euclidean perpendicular bisectors of $\overline{AZ\prime}$ and $\overline{AB}$ to find the desired center C for a Euclidean circle through A and B and $Z\prime$. Since this arc contains A and $Z\prime$, it must contain an elliptic line. Thus, that arc contains the third side of the elliptic triangle.

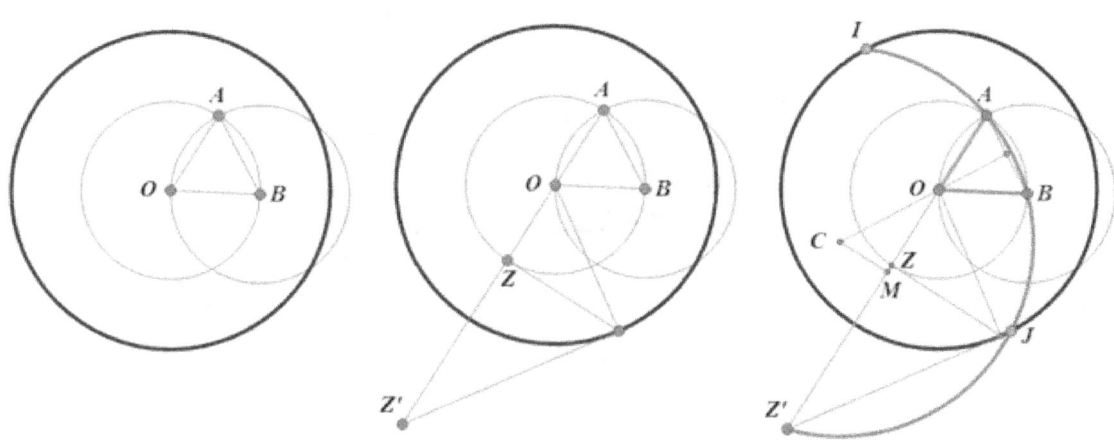

Is the elliptic triangle ABC regular? There's an excellent reason for the reader to find before the next sentence.

We know angle AOB is 60 since its sides are both elliptic and Euclidean. There's no way the other two elliptic angles can be 60 because that would be a Euclidean angle sum for an elliptic triangle. Of course, other reasons exist, too. But that was a fast one! A glance confirms that the Euclidean 60 degree angle at A does not have sides tangent to the elliptic angle OAB. It is interesting to note that a regular Euclidean skeleton does not automatically grant regularity to the elliptic object built from it.

3.13   Congruent triangles in the elliptic disk

Given elliptic lines $\overleftrightarrow{EA}$ and $\overleftrightarrow{EB}$ with Euclidean centers D and C respectively and lines $\overleftrightarrow{EO}$ and $\overleftrightarrow{DO}$ perpendicular, prove $\triangle ABE \cong \triangle ABF$.

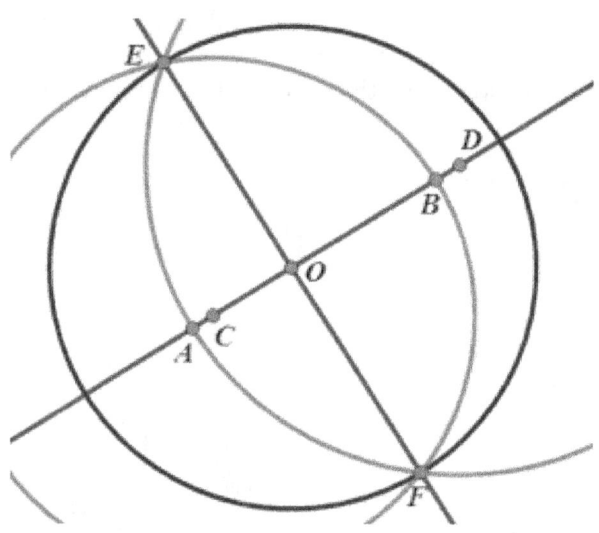

If the reader is thinking AAA, we are ready to go. Here's an acceptable answer.

Elliptic angles AEB and AFB are congruent by reflection across line $\overleftrightarrow{AD}$. The angles with sides on line $\overleftrightarrow{AD}$ are all 90 because the Euclidean centers are on that extended Euclidean diameter. So, the triangles are congruent by AAA.

3.14 Elliptic Geometry topics to know

Elliptic geometry rules, especially which negation of parallel postulate is in effect.

Construction of elliptic lines starting with two points on boundary.

Construction of elliptic lines starting with two points inside disk.

Area formula for lune and elliptic triangle.

Angle sum of elliptic triangle.

Elliptic poles and polars.

Constructions of all right-angled elliptic triangles.

Spherical distance formula using latitude and longitude.

Spherical distance formula applied to a 90, 90, A triangle.

Elliptic lines from Reuleaux triangle.

Elliptic squaring the circle

3.15 Elliptic geometry homework

(1) Would it be possible to construct an elliptic quadrilateral with opposite sides parallel? Think of parallel sides meaning the lines containing those sides do not intersect. Support your answer.

(2) Given two perpendicular diameters inside the elliptic disk, construct an isosceles elliptic triangle with two sides lying on these diameters.

(3) Suppose elliptic triangle ABC has all three right angles. Describe the polars for A, B, and C.

(4) Construct a lune with area $\frac{\pi}{2}$.

(5) Prove that we cannot have an elliptic trapezoid.

(6) Is the entire boundary circle an elliptic line? Explain why or why not.

(7) Given perpendicular diameters, $\angle HAE \cong \angle GFE$ and that elliptic lines $\overleftrightarrow{HF}$ and $\overleftrightarrow{GA}$ have Euclidean centers D and B respectively, all as pictured, prove $\triangle AHE \cong \triangle FGE$.

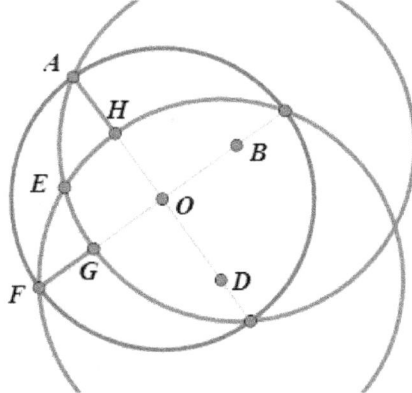

(8) Construct a pair of elliptic lines which meet at a 60 degree angle, at least one of which is not a diameter.

(9) Suppose elliptic quadrilateral $ABCD$ has right angles at A, B and C. What must be true about angle D? Construct such a quadrilateral.

(10) Fill out this table of geometric properties

	Hyperbolic	Euclidean	Elliptic
Sum of angles of triangle			
types of parallel lines			
number of sides of all right angled polygon			
area of triangle			

Problems 11 – 13 have appeared before in Euclidean and/or hyperbolic homework. Perhaps the strategies used before will work again.

(11) Prove that adjacent elliptic angles with exterior sides on an elliptic line are supplementary.

(12) Let angle ABD be an exterior angle of elliptic triangle ABC. Find the formula

for the measure of angle ABD in terms of angles A and C.

(13) Prove that if triangles ABC and DEF have corresponding angles congruent, then the triangles are congruent. (That means we are proving AAA in elliptic geometry.)

(14) What happens to all those nice properties of Euclidean parallelograms when we try to use them in elliptic geometry?

(15) The lune pictured seems to contradict the first axiom, that two points determine a line since we have two distinct elliptic lines through two distinct Euclidean points A and B. How do we avoid this contradiction?

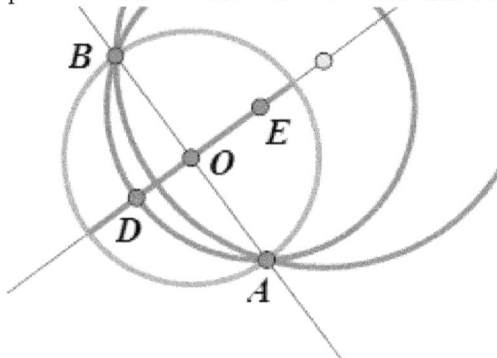

(16) In the figure for Problem **15**, prove angle BDO is 90 a right angle.

(17) Construct an elliptic triangle with exactly one right angle. Then construct an elliptic triangle with exactly two right angles. Finally, construct an elliptic triangle with 3 right angles.

(18) If you did not do this in problem **17**, construct a 90/90/90 elliptic triangle using only two radii of circle O.

(19) There's only one other elliptic polygon, besides the triangle, with all right angles. Construct an example of such a polygon.

(20) Prove the area formula for a lune on a sphere with radius r.

(21) Given Euclidean square $ECFD$ and angle DAO congruent to angle CBO. Prove elliptic triangle AOD congruent to triangle BOC.

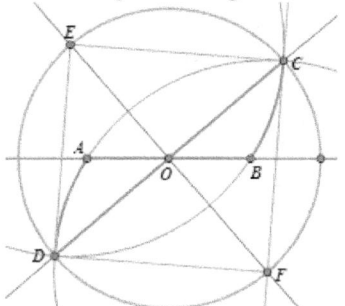

(22) Prove that the Reuleaux triangle we used as a regular elliptic triangle has 120 degrees per angle.

(23) Prove the $R^2 = 2r^2$ in our proof of the Reuleaux triangle construction.

(24) Construct an elliptic triangle with 135 degrees per vertex.

(25) Where are the poles for all diameters as polars? Support.

(26) Construct a regular elliptic quadrilateral.

(27) Given E the Euclidean center of arc DIC and F the Euclidean center of DHC, prove triangle HID congruent to triangle IHC.

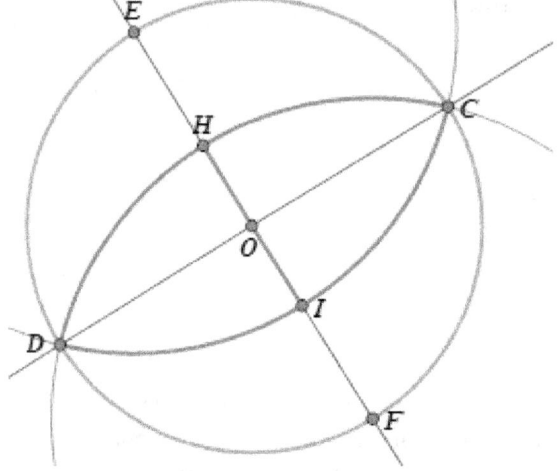

(28) Given Euclidean rectangle $ABCD$ with elliptic angle $AEB = 90$ degrees, prove elliptic angles ABE and BAE are obtuse.

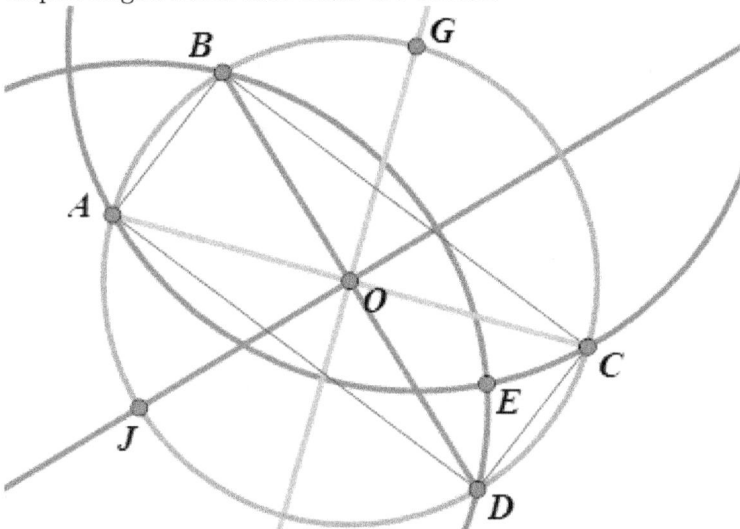

(29) Give yourself a point P, other than O, inside circle O. Construct the polar for pole P.

(30) What is the polar for pole O?

(31) Give yourself a point P on circle O. What is the polar for point P?

(32) In the process of squaring the elliptic circle, we need to construct a radius with a specific elliptic size. Kyle and Noah used an elliptic triangle with a vertex at O and two 90 angles positioned as pictured. The Euclidean center of the elliptic line is D. Prove $\cos A = \cos a$.

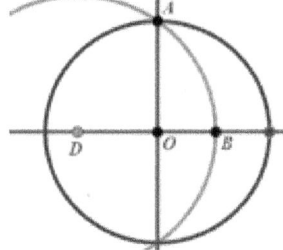

(33) We may use elliptic geometry to prove the independence of the parallel postulate from the other 4 axioms and isometries. Do so.

(34) Construct the Pi Mu Epsilon anniversary button pictured at the start of the Elliptic Geometry chapter.

(35) Construct the elliptic center of an elliptic circle.

(36) Prove that the elliptic length of a radius of the disk is $\frac{\pi}{2}$.

Chapter 4

For professors only

That title guarantees the students will at least read this section of the book. In this section, we discuss how to use this book as a text, a supplement to a text and as a source of undergraduate research questions.

We have an enormous source of questions when we have students familiar with compass and straightedge because we can request the construction of a particular (constructible) geometric object. We can also supply a construction and ask for its proof. While these are not the deepest questions in the course, they fill an essential need: quick assessment applicable to any week of the course.

Watching students work their drawing tools during class is a gas. Some will switch colors and provide a clean, brilliant construction. Others will draw each arc, circle and line four times each until their paper is a blur of imprecisions. A few will work in miniature, creating an exquisite construction which fits on a postage stamp. The professor can model good construction habits but that alone will not cure some students of bad habits. The personal touch can help a little bit.

The hassles with bad construction habits remind me exactly of bad solving habits or foggy use of the Chain Rule: when a student is not sure what to do, the drawing reflects their doubts. This book's lack of internet support places the students in unfamiliar territory without a map. The careful professor has to model the surety of thought we get from thinking about the task at hand and from experience. I have seen enormous improvement from week three to week ten in many students.

This is one of those courses where a successful student might have some clunky work from early in the course which the student could easily perform later. How the professor handles such gains in expertise is up to the professor, of course. Remember, besides the geometry content, the construction ability grows with the course, too.

I have mentioned *SketchPad* throughout the book. Of course any dynamic geometry software will work just fine. I have found my students develop skill with both the computer constructions and the physical compass and straightedge. Often, the physical tools work more quickly and they have the advantage of independence from a computer.

For assessment, I often create test problems where a student finishes a construction. This way, I can make sure that the work will fit in the space provided. I can also give a situation where very little needs to be done, IF the student understands the given. I gave a problem where a single segment was all that was necessary to construct and students went the long way around or missed the idea completely, while others smiled, performed the move and went to the next problem. Such thinking reminds me of Green's Theorem where one option works out much faster than the other integral.

The internet does indeed have constructions, video instructions and dependable collections of math ideas. All these tend to cover important, general ideas. So far, my students have had little success in collecting pages of notes from various websites to make this course easier. My most successful students always learn the course material, get good with constructions and end up thinking of their own solutions. I have taught from this book online, making up dozens of problems each week. Even with all that access and time, my successful students get with the program quickly, deciding that learning is faster than Googling.

The bright but inflexible students find this course maddening because they can learn all the definitions and theorems and construction moves and still freeze up on the open-ended problems. My only advice for them is experience. Feeding them a few step-by-step solutions, with the ultimate goal kept in mind, seems to help a little. But there is no substitute for studying the process, working through the steps. A person whose success has relied on giving back what has been given really is in trouble in this course, which starts with the given and seeks a specific conclusion. Giving back the given will not help at all!

The professor may get frustrated, too, when a student turns in a complicated, perfect construction which has been reproduced from memory but has little application to the given problem. The professor has to model, again, the solution process.

If the class has solved a problem together, with a few digressions or ideas which did not work out, my students have always appreciated a summary at the end of the problem. Such thumbnails get to be satisfying on their own as the course progresses because we develop a verbal short-hand. Also, the students recognize the careful use of notation to stand for complicated ideas. I love it when I can write, "construct the inverse of A across circle O" without having to write the steps for that construction because nobody needs those steps anymore. That's when I know my students are where I need them to be.

Sticking with constructions gives us a direct connection to mathematical work which used to be an essential part of an education. Spherical trigonometry, along with the development of accurate clocks, enabled people to navigate the globe at a time when finding specific places was a big deal. Even in this book, the non-Euclidean trigonometry gets little attention. But at least we construct with formulas a little. Our results indicate that more awaits us: clever construction short cuts, special cases with physical applications, design ideas for artists and more. If we can get our students to see geometry as a living, useful discipline, we will have done them a service.

Undergraduate research

Every autumn, the Van Andel Institute in Grand Rapids hosts an undergraduate science and math research symposium with hundreds of posters from the west half of the lower peninsula of Michigan. Every autumn, only one or two math posters appear while biology, chemistry and environmental science dominate with hundreds of posters. The students whose work fills these pages all took their turn being one of those few undergraduate mathematicians. They would have appreciated some company.

Although this book has been written primarily as a supplement to a college geometry course with emphases on constructions and proofs, the book also serves those professors who seek sources of open problems within reach of undergraduate students. Thanks to the generations of geometers who did not publish all their little tricks, there is room for exploration in geometry. Goodman-Strauss suspects that geometers must have known some of his hyperbolic construction moves and I heartily agree. Many of the non-Euclidean constructions in this book are too quick to never have been seen before. This might be why geometers did not publish the moves - they are not big results.

One source of ideas we have seen starts with nifty Euclidean results and then asks what happens in the two non-Euclidean geometries. We follow Bolyai's imagination when we think like this because he boldly claimed squaring the circle for hyperbolic geometry well before the Euclidean geometers had resolved the ancient question. He took an old Euclidean chestnut and cracked it open in a bent space. This trick worked for my students several times.

And it is true, having the adjustable angle sums of triangles in hyperbolic and elliptic give us a rich source of options. What is also fascinating are the times when those options do not help at all, when the bent space constructions do not work out so nicely.

As an example, I want to relate the development of Jillian's theorem, the constructible, regular, all-right angled hyperbolic polygons have the same number of sides as the constructible regular Euclidean polygons. In the summer of that work, we had been trying to construct an all-right angled regular hyperbolic heptagon because we had so many tools for getting perpendicular hyperbolic lines. I can still recall clicking and dragging the last endpoints. We could see where they needed to be, yet we could not find a way to get those points to those locations by construction.

Eventually, we saw that we had to know something ahead of time. We needed a plan. Jillian worked on constructions, building from Euclidean polygons to hyperbolic and eventually we realized that the structures go both ways: we could build hyperbolic from Euclidean and Euclidean from hyperbolic. Then her theorem made sense. We had banged against the wall enough to figure out that the wall was not a door.

So, remember, an open problem might turn out to be impossible, which in itself is a nifty result.

Remember that undergraduate research publications do not impress Provosts and Rank and Tenure Committees as much as publications in professional journals. Before diving into the bent spaces with some undergraduate students, the untenured professor should see how other such activities were received at his or her school.

Student researchers in this book

The student researchers in this book were students at Aquinas College and all had their work funded through the Mohler-Thompson Summer Research Program which Dr. Beth Jensen administers at Aquinas College. The author and the students wish to thank the Mohler family and Ted Thompson for their vision and generosity. Having the opportunity to immerse ourselves in geometry and to add to the understanding of important problems really is beyond price.

Jillian Duffey was Jillian Russo in the summer of 2009 when we worked on hyperbolic polygonal spirals together. She is teaching for an accelerated math program at the time of this writing. Her paper, "Hyperbolic polygonal spirals," is in Volume 11, Issue 2 of the *Rose-Hulman Undergraduate Mathematics Journal.*

Nathan Poirier and I worked together in the summer of 2010 but it took almost two years for the Alhazen result to evolve and to appear in print. Look for "Alhazen's problem in hyperbolic geometry" in Volume 5, No. 3 of *Involve.* Nathan finished his Master's Degree in mathematics in 2014, Western Michigan University.

Megan Ternes has her paper, "Tangent circles in hyperbolic space," in Volume 14, Issue 1 of the *Rose-Hulman Undergraduate Mathematics Journal.* She has two nice short-cuts for isosceles hyperbolic triangles which were too situational for this book. Megan was voted Outstanding Senior at Aquinas College when she graduated.

Noah Davis finished his research in 2012 and he and I tinkered with the paper for a while. His work was accepted at the *Rose-Hulman Undergraduate Mathematics Journal,* the Spring 2014 issue. All the details on squaring the circle in the Poincaré disk are there.

Kyle Jansens combined with Noah Davis to finish the ancient story of squaring the circle by figuring out how to square the circle in elliptic geometry in 2014. They feel safe about their results because they have verified their measurements using double integrals for surface area. They are preparing an article summarizing the Euclidean, hyperbolic and elliptic situations at the time of this writing.

Chapter 5

End Materials

5.1 Axiomatic Systems

An axiomatic system is a set of axioms, along with definitions and terms which are designated as undefined. The undefined terms can appear in the axioms and the definitions. Any other words used in the axioms and definitions take their usual meaning. Axioms are statements, that is, declarative sentences whose truth value can be figured out, when there is a context for these axioms.

We will now examine a particularly simple axiomatic system in order to see how the above paragraph works.

Axiom 1   Each tree has at least one squirrel.

Axiom 2   A squirrel can have at most one tree.

Axiom 3   There exists a tree.

Undefined terms: tree, have, squirrel.

There are several important things to observe about this axiomatic system. Since the verb "have" is undefined, the "has" in Axiom 1 is also undefined. This means that the possession usually associated with this verb is not necessarily involved. "Have" could mean a link between two objects, it could mean a directed relationship, like the subject is on top of the object of the verb. Mathematicians usually try to choose their undefined terms so that usage matches the word; it would be silly to write the verb "have" and build an axiomatic system where the verb clearly means "is greater than." It must be emphasized, however, that a mathematician is free to use whatever undefined terms he or she wants.

Similarly, the axioms have nothing to do with biological organisms. A squirrel could be almost any kind of noun. Since squirrel is used as a noun, then we could not expect squirrel to mean, "hit by a bus" or "throw that brick at me." So, undefined terms are not granted specific meaning; but we play fair – nouns are nouns and verbs are verbs.

An axiomatic system can have an interpretation. This is when we assign meanings to the undefined terms. Forming an interpretation usually leads to building a model, a mathematical contraption in which each axiom is true. We will now perform these two steps in our example.

Let's use the symbol £ to stand for a squirrel. We will use the symbol ¥ to stand for a tree. Since a squirrel can have a tree and a tree can have a squirrel, we will use the symbol of a single dash to stand for the verb to have. We may now write down a model:

£ £-¥

Our model satisfies Axiom 3 because it contains a tree. Axiom 1 forces this tree to have a squirrel, and the tree does indeed have a squirrel. This squirrel has only one tree. The other squirrel does not have a tree. These are the only two squirrels, so Axiom 2 is also true. Thus we have verified that we do indeed have a model of the axiomatic system.

The big point of an axiomatic system is to prove theorems. A proof is a psychological device meant to convince an educated reader that a theorem is true. Our little axiomatic system has an obvious theorem and we saw it in action in building our model.

Theorem: There exists a squirrel.

Proof: Axiom 3 gives us a tree. Axiom 1 gives us a squirrel for that tree.
■

This was a very good proof because we did not have to resort to our model – we reasoned from the axioms only. In *Geometry by Construction*, we have some rich axiomatic systems with axioms and definitions that interact to imply many theorems. The wording gets complicated and so we work with our models in order to see what we are doing. Our theorems are just as sound as if we had worked without an interpretation and without a model. Many college geometry courses spend a significant amount of time developing theorems in this way. This is called formal or axiomatic geometry. Its historical context is valuable because this was how non-Euclidean geometries were first seen. Mathematicians like Saccheri, Lambert, Gauss, Bolyai, and Lobachevsky were working formally in the 1800's, working blind because they were leaving the term "line" undefined. Their intellectual achievements motivated Poincaré to create his disk.

We get more from thinking about geometries as axiomatic systems than creating hyperbolic geometry. As mentioned in the course, the hyperbolic model proved that the parallel postulate is independent of the other axioms and isometries. This means that the parallel postulate cannot be proved from the other axioms and isometries. Proving that something cannot be proved is itself an unexpected intellectual accomplishment; such a proof takes on enormous historical importance when the proof addresses the status of a big target, like the Parallel Postulate.

To finish our tour of the basics of axiomatic systems, we will show that our Axiom 1 is independent of the other two axioms. We already have a model in which all three axioms are true. Now we need to find an interpretation and model in which the negation of Axiom 1, along with Axioms 2 and 3, are true.

The negation of Axiom 1 is There exists a tree that has no squirrel. Note the use of the same verb as the original axiom. We could have used a synonym, like there exists a tree without a squirrel, but we would have been making the mistake of using the usual English meaning of "has" when it was meant to remain undefined. We will use the same interpretation as before. Our model can now consist of

¥.

Yes, a single tree is all we need. The negation of Axiom 1 is true: we have a tree that has no squirrel. Axiom 2 is vacuously true: every squirrel does indeed have a tree. Yes, there are no squirrels. But nowhere is the existence of a squirrel required. Axiom 3 only requires the existence of a tree, not a squirrel.

A funny thing has happened in our second axiomatic system: the negated Axiom 1 implies Axiom 3 as a theorem. Since we were only worried about Axiom 1 being independent, this does not worry us. If, however, we were to study whether Axiom 3 were independent of the negated Axiom 1, we would find dependence.

So, we have a model in which all three axioms are true and we have a model in which the negated Axiom 1 is true along with Axioms 2 and 3. Then Axiom 1 must be independent. This is so because, if we could prove Axiom 1 from the other 2 axioms, our second model would have Axiom 1 and its negation true at the same time. Our second model would contain an obvious contradiction and our model does not.

This process is the standard way to prove that a statement cannot be proved from some other statements

5.2   References

H. S. M. Coxeter, *Introduction to geometry,* John Wiley and Sons, 1961.

J. Grey, *Worlds out of nothing: a course in the history of geometry in the 19th century,* Springer, 2007.

J. Grey, *Janós Bolyai, Non-Euclidean Geometry and the Nature of Space,* Burndy Library, 2004.

C. Goodman-Strauss, Compass and straightedge in the Poincaré disk, *American Mathematical Monthly,* **108** 1, 2001.

J. H. Conway and R. Guy, *The book of numbers,* Springer, 1996.

I. Niven, *Irrational numbers,* The Carus Mathematical Monographs, 1956.

W. Jagy, Squaring circles in the hyperbolic plane, *The Mathematical Intelligencer* 17 1995.

Student researchers

Noah Davis, Squaring the circle in the hyperbolic disk, *Rose-Hulman Undergraduate Math Journal,* **15**, 1, 2014.

Jillian Duffey (Russo), Hyperbolic polygonal spirals, *Rose-Hulman Undergraduate Math Journal,* **11**, 2, 2010.

Megan Ternes, Tangent circles in the hyperbolic disk, *Rose-Hulman Undergraduate Math Journal,* **14**, 1, 2013.

Nathan Poirier, Alhazen's hyperbolic billiard problem, *Involve,* **5**, 3, 2012.

Kyle Jansens and Noah Davis, Do-It-Yourself: Squaring the circle, to appear.

5.3 Hints to Selected Problems

Section 1 Euclidean Geometry

1. The Law of Sines will give useful ratios.

3. $\overline{CJ}$ is an altitude of triangle ABC. Congruent triangles have the same areas.

7. Find the center first.

17. Ceva's Theorem.

22. 145cm

28. Inversion preserves incidence. Use incidence of existing circles to get three points outside circle on the desired circle.

Section 2 Hyperbolic Geometry

6, 7, 8. We have to use tangent lines to measure angles.

9. The only worry here is placement. If we choose our starting points too close to the center O, then the Euclidean circles get huge. Too close to the boundary and the fourth vertex might not fit.

15. Once the given is made, the answer requires straight edge only.

19. The points are connected, so this is some sort of hyperbolic curve.

20. See Problem 15.

21. That theorem is false in hyperbolic geometry. Many examples, like the sum of the angles of a triangle is 180 degrees.

24. See Problem 21.

27. We cannot prove the negation of the Parallel Postulate because the postulate is independent of the axioms and isometries. (We merely verify the negation is true in our model.)

31. The hyperbolic center is on the boundary, where the two circles intersect. (So the hyperbolic center is not a hyperbolic point.)

Section 3 Elliptic Geometry

11. Use tangent lines.

19. The lune

20. Look down on the globe from above the North Pole.

22. Use tangent lines.

30. The boundary circle.

32. Use our only elliptic trig formula.

36. Use our only elliptic trig formula on a 90/90/90 quarter circle as an elliptic triangle.

Index

www.ingramcontent.com/pod-product-compliance
Lightning Source LLC
Chambersburg PA
CBHW081128170526

45165CB00008B/2587